86 Structure and Bonding

Springer

Berlin
Heidelberg
New York
Barcelona
Budapest
Hong Kong
London
Milan
Paris
Santa Clara
Singapore
Tokyo

Atoms and Molecules in Intense Fields

Volume Editors: L. S. Cederbaum,
K. C. Kulander, N. H. March

With contributions by
L. S. Cederbaum, K. Codling, L. J. Frasinski,
H. Friedrich, T. F. Gallagher, K. C. Kulander,
N. H. March, K. J. Schafer, P. S. Schmelcher

With 63 Figures and 1 Table

Springer

In references Structure and Bonding is abbreviated Struct. Bonding and is cited as a journal.

Springer WWW home page: http://www.springer.de

ISSN 0081-5993

ISBN 3-540-61015-4 Springer-Verlag Berlin Heidelberg New York

Typesetting: Macmillan India Ltd., Bangalore-25, India
SPIN: 10509022 66/3020 - 5 4 3 2 1 0 Printed on acid-free paper

Volume Editors

Professor N. H. March
University of Oxford, Theoretical Chemistry Department
South Parks Road, Oxford OX1 3UB / Great Britain

Professor Dr. K. C. Kulander
Physics Directorate, Lawrence Livermore National Laboratory
Livermore, CA 94551 / USA

Professor L. S. Cederbaum
Universität Heidelberg, Theoretische Chemie
Physikalisch-Chemisches Institut
Im Neuenheimer Feld 253, D-69120 Heidelberg / Germany

Editorial Board

Preface

The topic of atoms and molecules in intense external fields now embraces a wide interdisciplinary area of pure science and technology. As emphasized in the opening chapter of the present Volume, focused laser intensities exceeding 10^{15} W/cm^2 became available in the 1980s and with such intense laser electric fields a number of non-linear phenomena became accessible. A new generation of strong laboratory uniform magnetic fields of about 100 T and more, and of even more intense pulsed fields open new possibilities to study strong magnetic field effects in the laboratory. Particularly high magnetic fields (up to - 10^{8} T) play a decisive role in the physics and chemistry of the atmosphere of degenerate astrophysical objects like white dwarfs and neutron stars.

The nonlinear phenomena in intense magnetic and laser fields are given some prominence in this volume. Two chapters deal with multiphoton processes and time-dependent phenomena in atoms. In another chapter it is emphasized that the process of multi-electron dissociative ionization of molecules offers considerable challenges both for modeling and for the study of first-principles. The dynamics of molecules in such intense laser fields is an area of great interest, both at the time of writing and for future studies. In all these chapters the interplay between theory and experiment is demonstrated.

A relevant aspect of non-linear phenomena is chaotic dynamics. The way in which the theory of atoms in external fields impacts on the area of chaos is covered in some detail her. A chapter is devoted to the chaos induced by a time-dependent electric field with particular emphasis on chaotic aspects of scattering processes. A further chapter deals with the chaotic dynamics of neutral as well as of charged two-particle systems in uniform magnetic fields. Two chapters are concerned specifically with high magnetic field studies on atoms and molecules, especially the theoretical background. One is concerned with the classical and quantum phenomena arising in magnetic field owing to the nonseparability of the center of mass and internal motion. The appearence of new giant dipole states and self-ionization of charged systems are examples of such phenomena. Finally, there is a chapter which treats large numbers of electrons by exploiting semiclassical motivated by Bohr's Correspondence Principle. One area of considerable interest as this volume goes to press is whether multiply charged anions can be stabilized in magnetic fields now accessible in the laboratory. And will terrestrial chemistry be completely changed in the intense magnetic fields found at the surface of neutron stars?

We anticipate that the Volume should be useful for established research workers in the interdisciplinary area already referred to as well as for young scientists who are starting research in this field. Some parts of the book could also well prove useful in advanced courses for graduate students.

Our thanks are due to the publishers for their cooperation in all respects. One of us (NHM) wishes to thank Professor J. B. Goodenough for much support and encouragement and one of us (LSC) to Dr. P. Schmelcher for continuous collaboration. These have been instrumental in bringing this work to fruition.

Oxford, 1996 L. S. Cederbaum, K. C. Kulander, N. H. March

Table of Contents

Molecules in Intense Laser Fields:
an Experimental Viewpoint
K. Codling, L. J. Frasinski 1

Two Interacting Charged Particles in Strong Static Fields:
A Variety of Two-Body Phenomena
P. S. Schmelcher, L. S. Cederbaum 27

Semiclassical Theory of Atoms and Ions
in Intense External Fields
N. H. March 63

Field Induced Chaos and Chaotic Scattering
H. Friedrich 97

Microwave Multiphoton Excitation and Ionization
T. F. Gallagher 125

Time-Dependent Calculations of Electron and Photon
Emission from an Atom in an Intense Laser Field
K. C. Kulander, K. J. Schafer 149

Author Index Volumes 1 - 86 173

Molecules in Intense Laser Fields: an Experimental Viewpoint

K. Codling and L. J. Frasinski

J J Thomson Physical Laboratory, The University of Reading, Whiteknights, Reading RG6 6AF, UK

When molecules are subjected to intense laser fields, they undergo multielectron dissociative ionization (MEDI), a process that is, as yet, little understood. This chapter gives a brief overview of some of the experimental techniques used to study the process. At present there are two models that attempt to explain the facts, inertial confinement (with rapid ionization at a critical inter-ion separation) and laser-induced stabilisation. In order to ascertain which, if either, of these models is correct, more definitive experiments are required, involving interferometric control of laser parameters, good laser characterisation, and a more careful analysis of the dissociation dynamics.

1 Introduction 2

2 Basic Experimental Techniques 3
2.1 A Femtosecond Laser System 3
2.2 Laser Diagnostics 5
2.3 Vacuum Techniques 5
2.4 Electron Energy Determination 6
2.5 Ion Energy Determination 6

3 The Covariance Mapping Technique 9
3.1 Ion-Ion Covariance Mapping 9
3.2 Three-Dimensional Covariance Mapping 13
3.3 Electron-Ion Covariance Mapping 15
3.4 Electron-Electron 20

4 Experiments with Laser Pulse Manipulation 21
4.1 Interferometric Shaping with a Single Laser 21
4.2 Two Colour Excitation 22

5 High Harmonic Generation 23

6 Conclusions 24

7 References 25

1 Introduction

The emergence in the late 1980s of chirped pulse amplification techniques [1] meant that it was now possible to produce focused laser intensities well in excess of 10^{15} W/cm^2. This is equivalent to a laser electric field approaching one atomic unit and it is perhaps not surprising, therefore, that conventional perturbation theory cannot be applied to the dynamics of atoms and molecules in such intense laser fields. In fact these fields 'dress' the electrons and nuclei on a timescale that is short compared to those of conventional atomic or molecular processes and new non-linear phenomena are observed.

Above about 10^{13} W/cm^2 atoms exhibit the phenomenon of above threshold ionization (ATI), where many more photons are absorbed than the minimum required to ionize the atom [2]. The electron kinetic energy spectrum (the photoelectron spectrum) consists of a number of peaks separated by the laser photon energy – see Fig. 1. In addition, high harmonics of the laser frequency are generated with efficiencies much larger than can be accounted for by lowest order perturbation theory [3]. Molecules also exhibit photoelectron peaks associated with the ATI process [4] and high harmonic generation has recently been observed [5]. However, due to the additional degrees of freedom, molecules show phenomena that cannot occur in atoms, namely above threshold

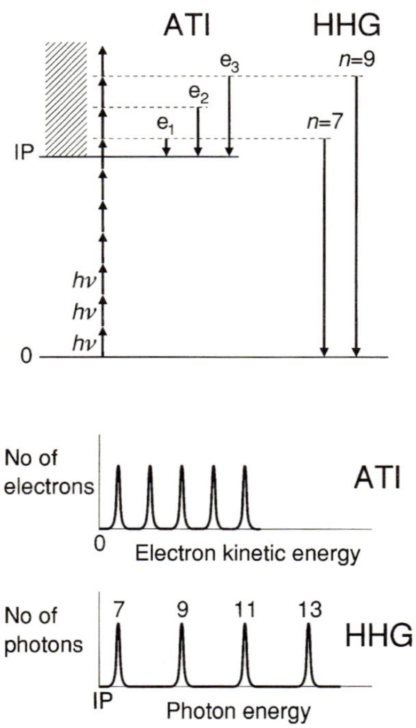

Fig. 1. Schematic diagram showing above threshold ionization (ATI) and high harmonic generation (HHG)

dissociation (ATD), or bond-softening [6] and bond hardening, or laser-induced stabilisation [7]. ATD has been observed in H_2, where it manifests itself in the proton energy spectrum as a series of peaks separated by half the photon energy. Moreover, when larger molecules such as N_2 and CO are exposed to laser intensities in excess of 10^{14} W/cm^2, multielectron dissociative ionization (MEDI) occurs and energetic fragment ions are observed [8]. Several mechanisms have been proposed to explain the MEDI process but at the present time Coulomb explosion combined with ionization enhancement at a critical internuclear distance appears to explain the bulk of the experimental data.

There are a number of excellent reviews on 'atoms in intense laser fields' [9]. Consequently, we will focus our attention on the experimental aspects of the behaviour of simple molecules in such fields. Most of the work to date has used a laser field of a single frequency, but experiments are presently underway where manipulation of dissociation pathways is being attempted by varying the phase between the fundamental and the second harmonic [10]. This simplest example of 'coherent control' of chemical reactions will be touched on briefly at the end of the paper.

2 Basic Experimental Techniques

MEDI is a complex process, where the timescales of dissociation and ionization generally overlap. In order to investigate the detailed dynamics of the process, three techniques are at our disposal, namely photoelectron spectroscopy (PES), ion mass spectrometry and conventional (photon) spectroscopy. As is the case in the single photon regime, much can be learnt using the first of these techniques, but this is the subject of another paper. High harmonic generation is an important aspect of intense laser physics, with potentially important application as a source of tunable, coherent, intense soft X radiation; this will be touched upon briefly in Sect. 5. We will concentrate on the photoions and describe how *coincidence* techniques (in this case covariance mapping) allow us to make definitive statements with regard to the experimental aspects of the MEDI process in simple molecules.

2.1 A Femtosecond Laser System

An example of a femtosecond laser system employing chirped pulse amplification is the one housed at the Laser Support Facility at the Rutherford Appleton Laboratory; a schematic layout is shown in Fig. 2. The system is essentially two-stage, with short, low energy pulses generated in the first stage and amplified in the second. To generate the pulses, a multi-line argon-ion laser with principal lines at 488 and 514 nm and 5 W continuous wave output is used to

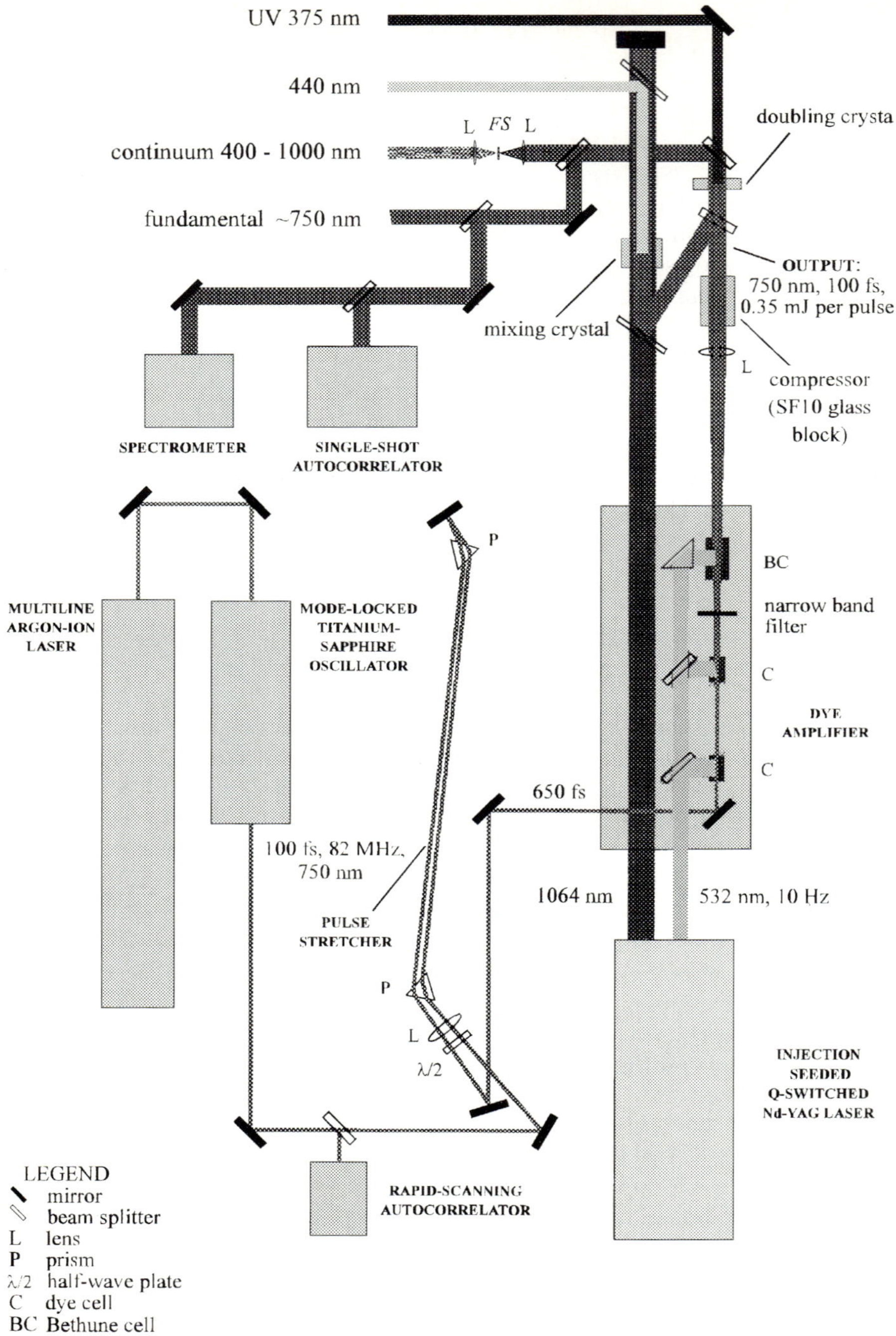

Fig. 2. The femtosecond laser system at the Laser Support Facility of Rutherford Appleton Laboratory

pump a passively mode-locked Ti:sapphire oscillator. This results in a train of narrow, intense pulses spaced by the cavity round-trip transit time, giving a repetition rate of 82 MHz. The pulses are variable in length from 100 to 300 fs, in wavelength from 730 to 800 nm with an energy output of about 2 nJ per pulse.

To overcome self-phase modulation and group velocity dispersion in the amplifier, one has to reduce the intensity by spreading the pulse in time. A suitable technique to achieve this is to employ a chirped pulse amplification and recompression scheme [11]. In the system described here, a pair of prisms is used to stretch the pulse to about 650 fs prior to amplification and to induce linear chirp. The pulses are amplified at 10 Hz through a three-stage dye amplifier pumped by an injection-seeded, frequency-doubled, Q-switched Nd:YAG laser which is synchronised to the Ti:sapphire oscillator. At 532 nm, the output is typically 250 mJ in 8 ns. The third stage, a Bethune cell, ensures a uniform gain distribution across the beam. A narrow-band filter inserted before the Bethune cell eliminates amplified spontaneous emission (ASE) from dye fluorescence. The stretched pulses are recompressed using a block of SF10 glass. Output pulses are typically 100 fs long, with an energy of 350 μJ and with less than 5% ASE.

2.2 Laser Diagnostics

It is important to have reliable laser diagnostics, preferably on a shot-to-shot basis; this is possible for a 10 Hz system. To characterise the temporal profile of the pulses, single-shot spectra and autocorrelation data can establish whether the laser pulses are Fourier-transform limited. For lasers with a sech^2 profile, a product $\Delta\nu\Delta t \approx 0.32$ should be achieved. It is also important to monitor the focal spot, its Airy disc and average intensity. Alternatively, a reasonable measure of the focused intensity can be obtained using Xe gas and the known threshold intensities for producing the various stages of ionization [12].

2.3 Vacuum Techniques

The experiments described in later sections require the use of ultrahigh vacuum (UHV) systems capable of achieving ultimate pressures in the range of 10^{-10} torr. One reason is that a tightly focused laser beam ionizes *all* molecules in the focal volume of about 10^{-10} cm^3. This can lead to space-charge effects, resulting in broadening and shifts of peaks in ion time-of-flight (TOF) spectra. To avoid such space-charge problems, beam pressures of $\sim 10^{-7}$ torr and ambient pressures of $\sim 10^{-8}$ torr are required. In fact for lasers giving focused intensities in excess of 10^{16} W/cm^2, ultimate pressures of better than 10^{-10} torr may be required.

A further reason to employ UHV techniques is the high ionization and fragmentation cross sections for hydrocarbons; H^+, C^+, CH^+, etc ion peaks can be particularly troublesome at the higher photon energies. One should therefore use well-trapped turbomolecular pumps or turbomolecular pumps with magnetic levitation, backed by a molecular drag pump, which itself is backed by an oil-free diaphragm pump.

2.4 Electron Energy Determination

Both electron and ion energies are usually determined by TOF spectroscopy. The reason is the high collection efficiency of the TOF systems. An ideal system to employ when accurate angular distributions are not required, or if coincidence measurements are not contemplated (Sect. 3.3), is the magnetic bottle spectrometer [13]. In this device, electrons originally emitted over 2π steradians from a small interaction volume are formed into a beam of half-angle 2°, i.e. they are parallelised. The instrument makes use of a magnetic field that reduces from 1 to 10^{-3} tesla as one goes away from the interaction region. Energy resolutions of 15 meV have been achieved. The device can also act as an electron-image magnifier.

2.5 Ion Energy Determination

When N_2 molecules are illuminated by an intense subpicosecond laser, high-energy fragment atomic ions N^+, N^{2+} and N^{3+} are observed, as well as thermal N_2^+ and N_2^{2+} ions. The first aim is to determine the numbers of these ions and their kinetic energies. This is usually achieved by ion TOF spectroscopy, as shown in Fig. 3. The laser light is focused by a doublet lens, L, onto an effusive beam of N_2 molecules. An electrostatic potential, U, forces the ions to travel along the upper drift tube to a pair of microchannel plates, MCP. The ion signal is transferred to a digital oscilloscope for manipulation.

Consider the creation and fragmentation of a particular transient ion, $[N_2^{3+}]$:

$$N_2 + nh\nu \rightarrow [N_2^{3+}] \rightarrow N^+ + N^{2+} + \text{kinetic energy.}$$

In this process the ions are ejected preferentially along the drift tube axis, that is along the E field of the highly linearly polarised laser. If one arranges the field-free (drift) region to be twice the length of the acceleration region, then the TOF is a linear function of the ion's initial momentum component along the drift tube axis [14]. (Incidentally, this arrangement also provides spatial focusing, that is it makes the TOF independent of the initial position of the ion. However, spatial focusing is superfluous in laser experiments as the ionisation volume is very small.)

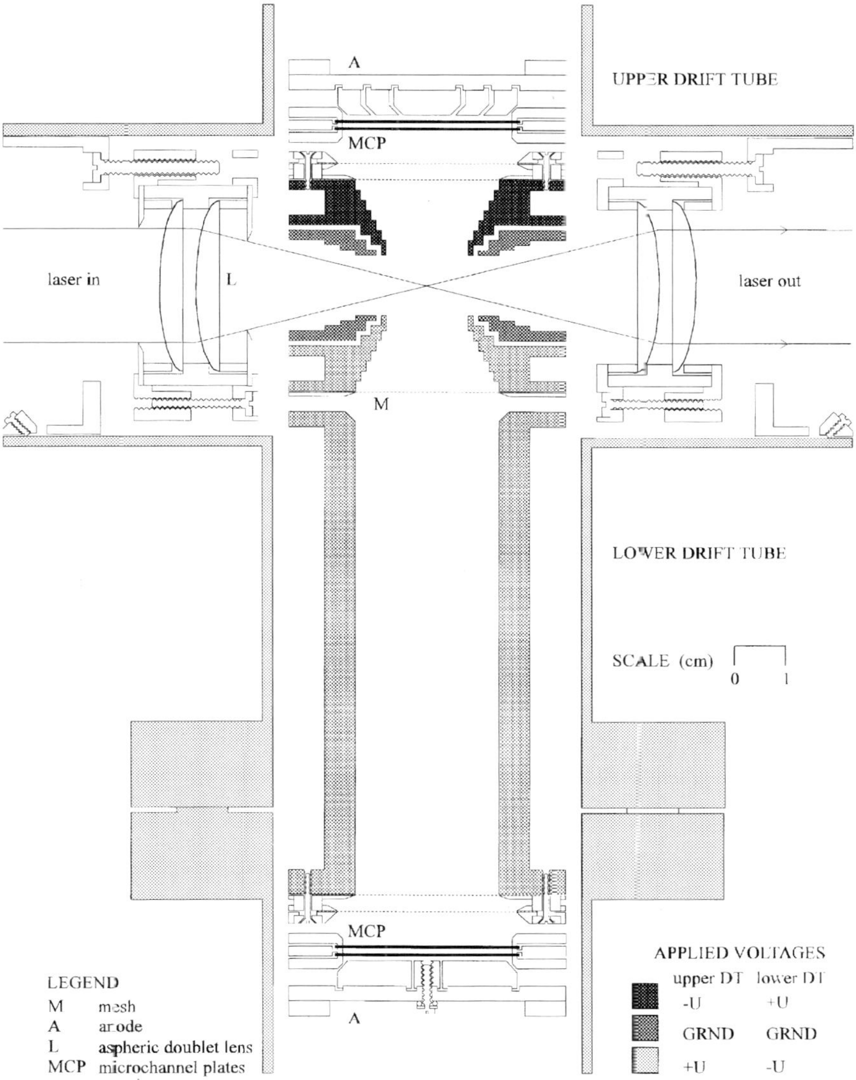

Fig. 3. A cross section of the ion-ion correlation analyser. The upper drift tube is used for covariance mapping, the lower one for high-resolution conventional TOF spectra

Normally, the field-free and acceleration regions are separated by a mesh. This mesh is often double to reduce the penetration of the acceleration field into the field-free region. It is possible to avoid using meshes by an appropriate electrode geometry. In this case the field varies smoothly along the drift tube axis and the acceleration and the field-free regions merge together. An example

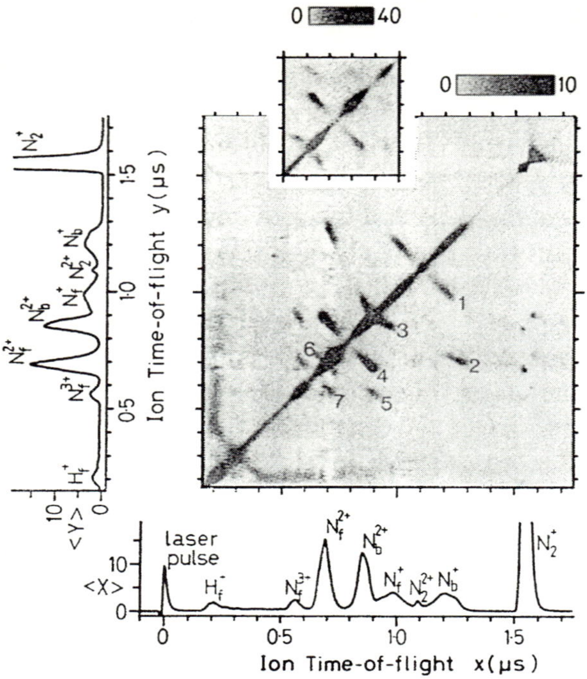

Fig. 4. Covariance map of N_2, with the conventional time-averaged TOF spectrum placed along the x and y axes. The correlated features 1–4 are discussed in the text. The inset shows the strong feature 4 using a larger range of grey scale

of such an arrangement is shown in Fig. 3, where the shape of the electrodes is optimized to give maximum energy dispersion in the ion TOF spectrum while ensuring that all ions, including the most energetic ones, are collected. The optimization was performed by numerical relaxation of the field distribution followed by ion trajectory simulation. The ion TOF in both mesh and meshless designs is given by the expression

$$t = A\left(\frac{m}{qU}\right)^{1/2} + B\left(\frac{mv_{\parallel}}{qU}\right) \tag{1}$$

valid for $mv_{\parallel}^2/2 < 0.1qU$.

Here A and B are instrumental constants, m and q are the mass and charge of the ion and $v_{\parallel}$ is the velocity component along the drift tube axis. Thermal ions arrive at the detector in a time given by the first term, 'forward' ions (labelled N_f^{2+}) with momentum $mv_{\parallel} < 0$ (i.e. ejected initially towards the detector) arrive at an earlier time and 'backward' ions (labelled N_b^{2+}) with $mv_{\parallel} > 0$ arrive at a later time. A typical low-resolution, time-averaged TOF spectrum of N_2 at the bottom of Fig. 4 shows the double-peaking that results.

We are now faced with a problem that cannot be resolved by an improvement in resolution. We cannot state categorically whether the N_f^{2+} ions originate from the process $[N_2^{3+}] \rightarrow N_b^+ + N_f^{2+}$ or from the process $[N_2^{2+}] \rightarrow N_b - N_f^{2+}$; the TOF system is incapable of detecting neutral fragments. One may be able to assign some of the N_f^{2+} ions to the first process if they have the same kinetic energy as the N_b^+ ions (conservation of linear momentum). However, what is required is a method of correlating the fragment ions unambiguously.

Frasinski et al. [15] devised a triple coincidence technique to study double ionization in the single photon regime. In these photoelectron-photoion-photoion coincidence experiments, the photoelectron provides the 'start' pulse to an analogue-to-digital converter and the two ions the 'stop' pulses and it was an easy matter to ensure that there was less than one molecular ion in the interaction region for the period of measurement of the ion energies, about 10 μs. In the multiphoton case, however, may ions are produced per laser pulse and this conventional approach cannot be employed.

3 The Covariance Mapping Technique

3.1 Ion-Ion Covariance Mapping

In 1989, Frasinski et al. [16] introduced a new triple coincidence technique called covariance mapping. A brief explanation of the technique can be given by considering again the process $[N_2^{3+}] \rightarrow N_b^+ + N_f^{2+}$ and denoting by t_1 and t_2 the TOFs of the N_f^{2+} and N_b^+ ions respectively. If an ion is detected at time t_1, then there is clearly an enhanced probability of detecting an ion at time t_2. When one calculates, over many laser pulses, the covariance between the two TOF points, one finds a positive value. Since one does not know ahead of time which ions in the TOF are correlated, one must calculate the covariance of each pair of TOF points and thereby obtain a two-dimensional map. The covariance $C(t_1, t_2)$ is given by the expression

$$C(t_1, t_2) = \langle X(t_1)\, Y(t_2) \rangle - \langle X(t_1) \rangle \langle Y(t_2) \rangle \tag{2}$$

where X and Y are the detector responses at time t and the angular brackets denote an average over many laser pulses.

In this case, $X(t)$ and $Y(t)$ are the *same set* of one-dimensional TOF spectra but in general they can be different, as in the electron-ion experiment discussed later. Figure 4 shows a typical map of N_2 taken with 10 000 laser pulses; it should be emphasised that a contribution to the covariance must be compared at each laser pulse. For clarity the conventional time-averaged ion TOF spectrum is shown along the x and y axes.

By creating such a map one can obtain all of the dissociation channels involving correlated charged fragments. One usually sees the different channels as pairs of separate features (islands) on the map, because the energies have a limited range. Momentum conservation and the TOF linearity with ion momentum ensures that the structure lies on a straight line with a slope given by the ratio of the two charges. Feature 1 involves $N_b^+ + N_f^+$, whereas the feature reflected in the strong 45° autocorrelation line involves $N_f^+ + N_b^+$. Feature 2 involves the channel $N_b^+ + N_f^{2+}$, whilst feature 3 involves $N_f^+ + N_b^{2+}$. The strongest feature, 4, is associated with the charge-symmetric channel $N_b^{2+} + N_f^{2+}$.

One sees that, even if the N_b^+ features 1 and 2 have *identical* energies and overlap completely in the one-dimensional ion TOF spectrum, covariance mapping allows one to apportion these N^+ ions to the parent molecular ion precursors. By scanning along the double feature, one can determine branching ratios (partial cross sections) for the various ion pairs. Since two ions must be involved if a process is to be seen on the map, whereas one ion is sufficient for it to appear on the one-dimensional TOF spectrum, one can gain information on processes involving neutral fragments by projecting features on the map onto the x axis and comparing them with the TOF spectrum.

It is perhaps worth noting that the distance from interaction region to detector in the upper drift tube is only 30 mm. This ensures 100% collection efficiency for even the most energetic of 'backward' ions; this is important if one wishes to correlate all the ions and obtain meaningful branching ratios. A TOF spectrum with higher resolution can be obtained by reversing the polarity on the electrodes and using the lower drift tube – see Fig. 3. A pinhole in front of the detector removes any effects due to angular distributions and allows an accurate measurement of ion energy.

An alternative approach [17] is to map the correlation coefficient $R(t_1, t_2) = C(t_1, t_2)/[C(t_1, t_1)\ C(t_2, t_2)]^{1/2}$. The point of using this correlation coefficient, which ranges from zero to unity, is that it emphasises the weaker channels. Of course, the signal-to-noise ratio is not enhanced using this approach and, moreover, branching ratios are not obtained. A more conventional triple coincidence technique has been used to correlate charged fragments in I_2 [18]. The pressure and laser intensity are adjusted to detect less than one ion per laser shot, on average. The digital oscilloscope triggers and transfers data to the computer only if iodine atomic ions are detected. The computer analyses each TOF spectrum and if two ions are detected, a point is plotted on the correlation map. Events with one or more than two ions are discarded. Iodine ions are detected in 5% of the laser shots and 1% were two-ion events. The results indicated that $[I_2^{4+}]$ ions were as likely to fragment charge-asymmetrically ($I^+ + I^{3+}$) as charge-symmetrically ($I^{2+} + I^{2+}$), when using a pulse length of 80 fs. However, other groups, using covariance mapping, find that fragmentation is predominantly charge-symmetric at larger pulse lengths [19, 20].

The power of covariance mapping will be illustrated by one example. When Lavancier et al. [21] studied the multiple ionization of CO at 305 nm, analysis of their conventional TOF spectra suggested that the process $[CO^{2+}] \rightarrow C^{2+} + O$ was the primary method of producing C^{2+} ions; there were no signs of the $C^{2+} + O^{+}$ channel. However, subsequent covariance maps showed clear evidence of the $C^{2+} + O^{+}$ channel [22]. Moreover, by comparing the covariance map with the concurrently-accumulated TOF spectrum (which contains ions from ion-neutral and ion-ion processes), the $C^{2+} + O$ channel was estimated to be less than 10% of the $C^{2+} + O^{+}$ channel. Subsequent inspection of the earlier TOF spectra showed that the O^{+} ions associated with this latter channel were hidden in the broad O^{+} structure associated with the $C^{+} + O^{+}$ channel.

One should remember that covariance mapping simply correlates the *final* ion states (ion pairs); intermediate states must be deduced from the final states or by appropriate modelling of the process. An illustration of the problem encountered is highlighted by an ongoing discussion as to whether the MEDI process is charge-symmetric or charge-asymmetric. Covariance mapping shows conclusively that the final products are charge-symmetric, i.e. their charges differ by no more than one elementary charge. (Such widening of the charge-symmetry definition is needed because a diatomic molecular ion with an odd charge clearly cannot fragment charge-symmetrically). However, when Hill et al. [23] studied N_2 at 193 nm using pulses of 10–15 ps duration, they explained their low energy N^{2+} and N^{3+} ions in terms of 'vertical' multiphoton excitation to asymmetric states $N + N^{2+}$ and $N + N^{3+}$. (Because covariance mapping was not used in this experiment, it is not known whether the N^{2+} and N^{3+} ions could be associated with the $N^{2+} + N^{2+}$ and $N^{3+} + N^{3+}$ channels.) This 'nascent' ionization can be followed by 'post-dissociation' ionization (PDI). In the former process a correlated neutral-ion pair is created almost immediately in a vertical transition; this might lead naturally to asymmetric fragmentation, for example $N + N^{3+}$, if a short laser pulse were used. However, if a longer pulse is employed, PDI produces a more symmetric outcome because of subsequent ionization of the neutral atom at large internuclear separation.

In their I_2 paper, Dietrich et al. [24] discuss the implication of these ideas more fully and suggest that their observation of asymmetric channel $I^{+} + I^{3+}$ and $I^{2+} + I^{4+}$ is consistent with a pulse length of 80 fs; they are observing nascent ion production. The observation of charge symmetry in I_2, N_2 and CO with pulses approaching 1 ps, they argue, is simply the result of PDI. On the other hand, covariance mapping has recently been applied to I_2 using pulses of 150 fs [19] and 200 fs [20] and the $I^{+} + I^{3+}$ channel is not observed.

Ion kinetic energy measurements can be used to assess the plausibility of PDI. If $N + N^{3+}$ were to be created in a vertical transition and the N atoms subsequently ionized at large internuclear separations, then the energies associated with the channels $N^{+} + N^{3+}$, $N^{2+} + N^{3+}$ and $N^{3+} + N^{3+}$ should be very little different. Experiments at 610 nm with 2 ps (i.e. long) pulses [17] give the following energies: $N^{+} + N^{3+} + 18$ eV (very weak); $N^{2+} + N^{3+} + 35$ eV; $N^{3+} + N^{3+} + 53$ eV. Of course, what is meant by large distances is a matter of

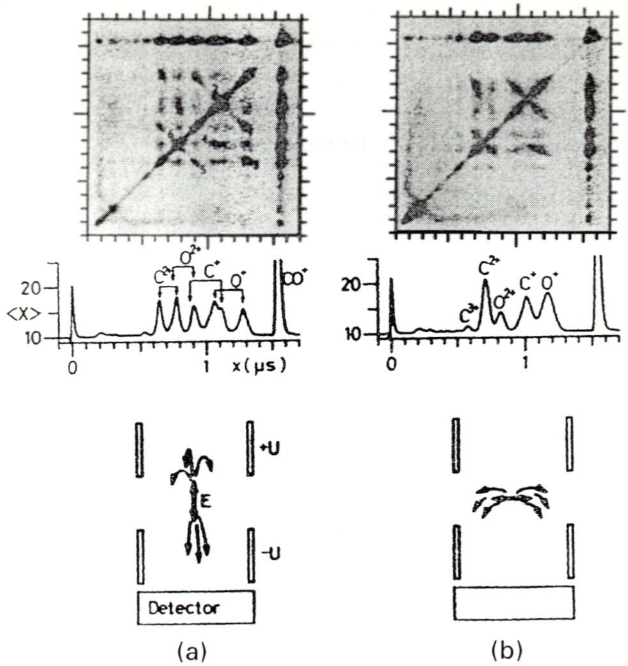

Fig. 5a, b. Covariance map of CO with the laser *E* field: **a** parallel to; **b** perpendicular to the drift tube axis. The features labelled 1–6 are discussed in the text

semantics but this sequence of energies could hardly be explained in terms of PDI. We will return briefly to this sequence of energies in Sect. 4.1 and discuss possible explanations for it.

Before leaving the subject of ion-ion covariance mapping, we comment on experiments measuring ion angular distributions. The very first experiment on the MEDI of N_2 noted the strong directionality of the fragment ion emission relative to the laser *E* field [8], but no quantitative measurements were made. Subsequently isoelectronic CO was studied at 600 nm [25] and an example of the covariance maps produced when the laser *E* field is (a) parallel to and (b) perpendicular to the drift tube axis is shown in Fig. 5.

Any feature on the map is, inevitably, a fold of the momentum distribution and angular distribution and in general it is impossible to unfold the two. However, in this case the momenta have a limited range and the angular distributions are highly peaked along the laser *E* field. This means that deconvolution is possible. In Fig. 5a the features are as follows: 1, $C_f^+ + O_b^+$; 2, $C_b^+ + O_f^+$; 3, $C_f^{2+} + O_b^+$; 4, $C_b^{2+} + O_f^+$; 5, $C_f^{2+} + O_b^{2+}$; 6, $C_b^{2+} + O_f^{2+}$. Having achieved the desired objective of isolating the various channels, it was possible to determine the kinetic energy releases and angular distributions associated with each. Analysis of the data showed that the angular distribution appeared to become more peaked as the stage of ionization increased.

The initial explanation of the strong directionality was in terms of the field ionization model and the molecule's elongated shape [26]. The potential difference created by the laser is largest when the molecule finds itself with its axis aligned with the laser electric field. The barrier to tunnelling is lower and the ionization easier. Since there is little time for molecular rotation, the ions are also ejected along the laser field.

Two recent publications, one on CO [27], the other on I_2 [28], explain the angular distribution in terms of laser-induced alignment. The laser polarises the molecule and induces a torque that forces the molecules to align with the *E* field. This has been studied experimentally by firing two laser pulses in quick succession at the molecular target, the pulses having orthogonal polarisations. It was concluded that both pulses interact essentially with all molecules, regardless of their initial orientation.

3.2 Three-Dimensional Covariance Mapping

We have seen, in the case of diatomic molecules, that two-dimensional covariance maps allow correlation of the ion pairs. The situation with regard to the multiple ionization of triatomic molecules is more complex. Similar two-dimensional maps serve to indicate correlations between pairs of ions but at high laser intensities it is probable that three ions will be produced. Although one might infer that three ions have been created simultaneously, the only sure way of confirming the creation and subsequent fragmentation of a triple ion is to use three-dimensional covariance mapping.

A two-dimensional map of N_2O taken at 600 nm is shown in Fig. 6, where one sees a weak feature of 45° in the top right hand corner of the map; this is associated with the two-body fragmentation $[N_2O^{2+}] \rightarrow NO^+ + N^+$. The features of interest here are the three strong features and their reflections in the 45° autocorrelation line. Their location on the map (i.e. on the TOF spectrum shown alongside) indicates that they must be associated with doubly charged ions. However, any analysis of the map must be somewhat ambiguous because one cannot differentiate the situation where pairs of ions are produced independently:

$$\begin{aligned} &[N_2O^{4+}] &&\rightarrow N^{2+} + N_m^{2+} + O \\ \text{and}\quad &[N_2O^{4+}] &&\rightarrow N^{2+} + N_m + O^{2+} \\ \text{and}\quad &[N_2O^{4+}] &&\rightarrow N + N_m^{2+} + O^{2+} \end{aligned}$$

from the situation where three ions are produced simultaneously

$$[N_2O^{6+}] \rightarrow N^{2+} + N_m^{2+} + O^{2+} .$$

(The subscript m denotes the middle atom of this linear triatomic molecule.)

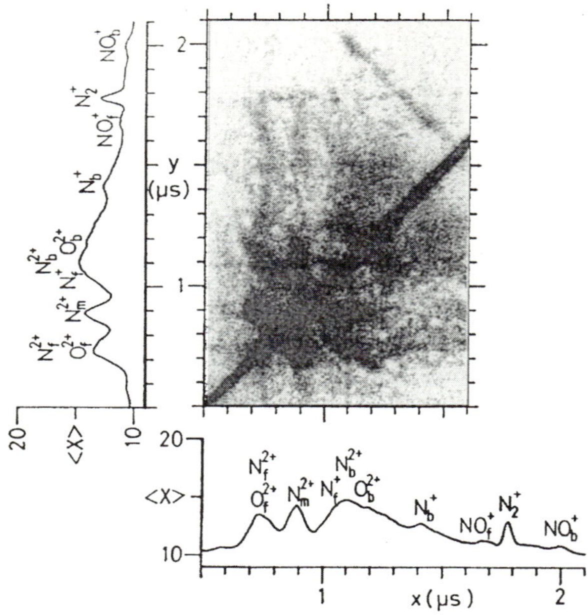

Fig. 6. Covariance map of N_2O, with the conventional ion TOF spectrum placed along the x and y axes

In order to remove this uncertainty, a three-dimensional map is required. In this case one is looking for correlations between three points (t_1, t_2, t_3) in the TOF spectrum. The covariance is given by

$$C(t_1, t_2, t_3) = \langle X(t_1) X(t_2) X(t_3) \rangle - \langle X(t_1) X(t_2) \rangle \langle X(t_3) \rangle - \langle X(t_1) X(t_3) \rangle \langle X(t_2) \rangle - \langle X(t_2) X(t_3) \rangle \langle X(t_1) \rangle + 2 \langle X(t_1) \rangle \langle X(t_2) \rangle \langle X(t_3) \rangle \,. \tag{3}$$

This expression must be evaluated at all times t_1, t_2 and t_3 in the TOF spectrum to produce a three-dimensional (cubic) map where one can imagine the TOF spectrum lying along the x, y and z axes [29]. Apart from the autocorrelation planes, and any statistical noise, only true triple fragment correlations will give real counts on the map. Six three-dimensional features should appear for each triple correlation; these occur at (t_1, t_2, t_3) (t_3, t_1, t_2), (t_2, t_3, t_1), (t_3, t_2, t_1), (t_1, t_3, t_2) and (t_2, t_1, t_3). If one looks down on the cube from above one sees the six features of Fig. 6. In order to locate one of these features, one takes consecutive slices of the cube. The signal-to-noise on these maps is worse than that shown in Fig. 4, reflecting a triple product of detector efficiency rather than a double product.

What is surprising is the lack of peaks in the TOF spectrum (and structure on the covariance map) associated with single ions. The total energy release (about 40 eV) suggests that, as the three atoms move apart and as the laser E field increases, there comes a point when the molecule is suddenly six-times ionized. One can speculate on the mechanism involved, but this interesting phenomenon deserves further study both experimentally and theoretically.

3.3 Electron-Ion Covariance Mapping

When even the simplest of diatomic molecules, H_2, is subjected to laser fields in the range 10^{12}–10^{14} W cm^{-2} the resulting dynamics can be extremely complex. A number of dissociation or ionization channels are accessed, each occurring with a branching ratio depending critically upon the laser wavelength, risetime and focused intensity [30]. For example, resonantly enhanced dissociative ionization can occur and transient bound states can be created by avoided crossings [31]. This bond softening may produce additional structure in both the photoelectron and photoion spectra. Lasers with short rise times may result in bond hardening, where the protons have higher energies than with a longer pulse [32]. If the laser rise time is reduced further, the two electrons may be ejected virtually instantaneously, with the electrons sharing the excess energy in a more well-defined way. The protons may show evidence of a Coulomb explosion, rather than bond-softening or bond-hardening. Add to these competing mechanisms the structure associated with ATI and the dynamic Stark shift and one sees that the analysis of conventional photoelectron spectra, even at modest laser intensities (up to 10^{14} W/cm^2), is by no means straightforward.

As the intensity rises well above 10^{14} W/cm^2, a molecule such as N_2 has many fragmentation pathways involving N^+, N^{2+} and N^{3+} fragment ions. The measured photoelectron spectrum will be the sum of contributions from all possible processes and, since energy shifts are almost inevitable, the conventional, time-averaged photoelectron spectrum may be impossible to analyse. In order to be able to associate each photoelectron peak with a parent or fragment ion, an electron-ion correlation experiment is required. In the case of electron-ion covariance, $X(t)$ in Eq. (3) is now the photoelectron TOF spectrum recorded *at the same time* as the ion TOF spectrum, $Y(t)$.

To date, there has been only one experiment using electron-ion covariance to study the MEDI of H_2 [33]. The experimental arrangement used is shown schematically in Fig. 7. A laser of 600 nm wavelength of 3 ps pulse length is focused to give an intensity of about 10^{15} W/cm^2. An electric field across the interaction region accelerates electrons towards the electron drift tube, ions towards the ion drift tube. As before, the charged particles are detected by microchannel plates, MCP. An axial magnetic field provided by field

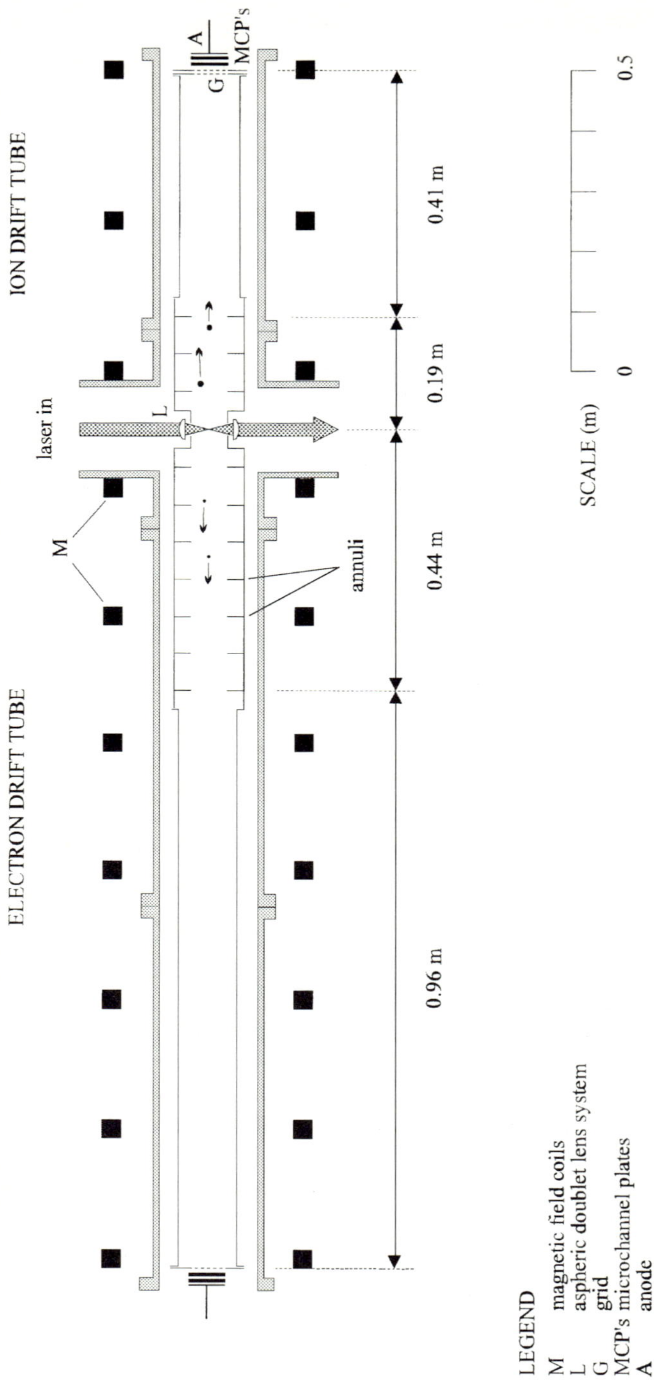

Fig. 7. Schematic diagram of the electron-ion correlation analyser

coils, M, ensures efficient collection of the electrons by guiding them to the detector.

Since the laser is highly linearly polarised with its E vector along the drift tube axes, the electrons are ejected initially towards or away from the electron detector. The ions are also ejected preferentially towards (H_f^+) or away from (H_b^+) the ion detector. The design of the double drift tube is such that the forward and backward electrons *and* forward and backward ions arrive at their respective detectors with a TOF, t, given by Eq. (1); the instrumental constants, A and B, differ for electrons and ions. Hence the electron and ion TOF spectra are symmetric about their respective zero energy TOFs.

Fluctuating laser intensity introduces false correlations on covariance maps. As the production of ions of all charge states increases with laser intensity, such fluctuations correlate erroneously every ion with every other ion. Fortunately, false correlations are proportional to the square of the sample pressure, while true ones are linearly proportional to the pressure. Therefore, at sufficiently low pressures the false correlations become negligible. Figure 8a shows an ion-ion covariance map of H_2 at an ambient pressure of 8×10^{-7} torr. The circular shape of features 1 and 2 suggest a problem with false correlations; this is confirmed by the observation of impossible correlations between H^+ and H_2^+. At the lower pressure of 2.4×10^{-8} torr, Fig. 8b, the spurious features have disappeared and the features at 45° ($t_x = -t_y$) are true correlations between two protons originating from the same molecule. The one-dimensional TOF spectrum shows three peaks, associated with protons with kinetic energies 0.35, 1.4 and 2.4 eV. One notes that these are separated by about 1 eV or half the photon energy and presumably indicate ATD or bond softening.

Figure 9 shows an electron-ion covariance map obtained at the same time as Fig. 8b. For convenience the time-averaged electron and ion TOF spectra are shown alongside the x and y axes. The electron TOF is virtually structureless, the slight forward-backward asymmetry and the peak at 292 ns being due to a small inhomogeneity in the extraction field. The forward electrons are unaffected and a kinetic energy scale is shown for both electrons and protons. In order to ensure that this lack of structure was not the result of poor instrumental resolution, an electron TOF spectrum of Xe was taken under identical conditions. A series of ATI peaks was observed in both the forward and backward direction, separated by about 2 eV.

Figure 9 therefore shows, at a glance, the photoelectron spectrum associated with thermal H_2^+ ions and with H^+ ions of all energies, that is with nondissociative and dissociative ionization. With the modest resolution and signal-to-noise ratio achieved in this first experiment, any vibrational structure associated with H_2^+ is not recognisable. Nor is there any structure associated with the H^+ ions. Nevertheless, in situations where the photoelectrons from the different ionization processes cover a similar energy range, as is the case here, the electron-ion covariance technique provides a powerful method of separating the various contributions.

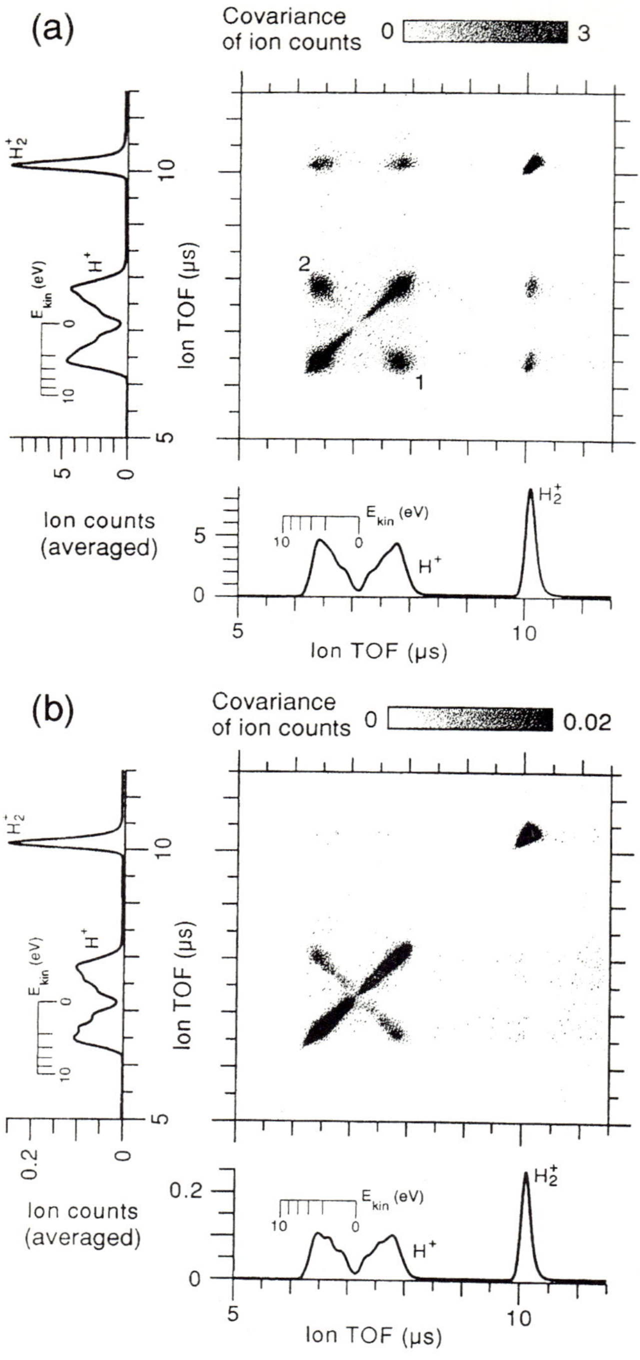

Fig. 8a, b. Ion-ion covariance map of H_2: **a** at 8×10^{-7} torr; **b** at 2.4×10^{-8} torr ambient pressure. The features are 1, $H_b^+ + H_f{}^+$; 2, $H_f^+ + H_b^+$. The three features at 45° ($t_x = t_y$) are parts of the autocorrelation line

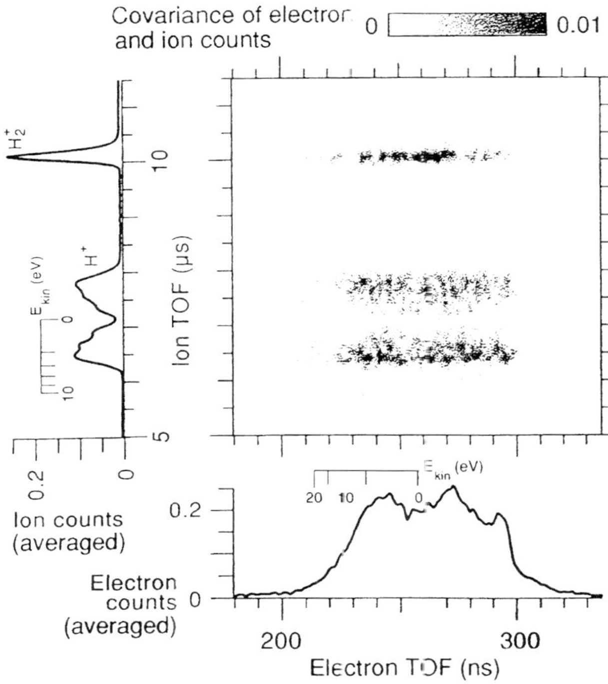

Fig. 9. Electron-ion covariance map of H_2 at 2.4×10^{-8} torr

The data represented by Fig. 9 can also be used to determine the relative probabilities of single and double ionization. The map registers only the $H^+ + H^+$ process, while the conventional TOF spectrum shows protons produced in both processes. By projecting the map features onto the horizontal axis and subtracting the result from the conventional TOF spectrum, one can isolate the $H + H^+$ contribution. In fact one sees that, in general, the low-energy H^+ ions are the result of the single ionization process and the high-energy H^+ ions the result of double ionization. More importantly, one can apportion the various H^+ ions to the two processes. It is interesting to note that recent work on ATD with ultrashort laser pulses (100–150 fs) at 769 nm indicates a similar behaviour [32]. Analysis of their conventional proton TOF spectra suggested that the low energy protons (less than 2 eV) were to be attributed to ATD via bond softening, whereas those protons with energies from 2 to 5 eV were to be associated with photoionization from laser-induced bound states of H_2^+; this ionization route produces two correlated protons. However, the need to conjecture about the particular route to proton emission can be removed by use of the electron-ion covariance technique.

3.4 Electron-Electron Covariance Mapping

Perhaps the most interesting application of electron-electron covariance mapping relates to the question of the major mode of multiple ionization of atoms and molecules. Luk et al. [34] studied the multiple ionization process in Xe using a laser of 193 nm wavelength and 10 ps pulse length and conventional ion TOF spectroscopy; they suggested that it was direct (a collective, instantaneous emission of many electrons). Lambropoulos [35] pointed out that, with a laser of such modest rise time, the ionization must proceed sequentially. In fact L'Huillier et al. [36] had also studied Xe at 532 nm and observed a 'knee' in the curve of log (ion counts) vs log (laser intensity) for Xe^{2+} that they attributed to direct double ionization.

More recent experiments by Walker et al. [37] using linearly and circularly polarised light and ion TOF spectroscopy confirmed the existence of the knee but, more importantly, electron-ion and electron-electron coincidence experiments were performed. Using the electron-ion coincidence technique, it was found that the great majority of the Xe^{2+} ions were produced in a sequential process. Moreover, direct double ionization was not observed in these electron-electron coincidence experiments. Charalambidis et al. [38] considered this long-standing question of direct double ionization in Xe and also concluded that it does not occur. That does not mean that correlated double emission cannot occur under appropriate conditions.

If these conditions are met, the two electrons from a direct ionization process will share the available excess energy in various ways, producing a continuous electron energy spectrum that would be hard to distinguish from background signal in a conventional electron TOF spectrum. However, if one produces an electron-electron covariance map, and if the two electrons are removed very rapidly (such that the direct process wins over the first step of the sequential process), one might expect to see a recognisable feature on the map. This is because the map is a momentum correlation map, and the sum of the squares of the momenta of the two electrons should be constant. An annular feature should therefore be observed on the map. In the H_2 experiment discussed earlier, the electron-electron map was unstructured and this was assumed to reflect the fact that the ionization process was sequential and not direct.

Recently Corkum [39] has pointed out that, during strong-field multiphoton ionization, an ejected electron has a significant probability of returning to the vicinity of the ion with high kinetic energy and scattering. High harmonic generation (see Sect. 5) and multiphoton two-electron ejection are both consequences of this electron-ion interaction. One interesting aspect, in the context of double ionization, is that the electrons in this internal 'e-2e' collision will be correlated, but it is not known how such a process would show up on an electron-electron covariance map. Such collisions are expected to occur for linearly polarised light, but for circularly polarised light the electron's trajectory is such that it never returns to the vicinity of the ion. It will certainly be interesting to see what happens in the case of a diatomic molecule exposed to a laser of very short pulse duration.

4 Experiments with Laser Pulse Manipulation

Under this heading are described two types of experiments where the laser pulse shape is modified, either by interferometric mixing of a single laser or by introducing a second laser with a different frequency, usually a second or third harmonic.

4.1 Interferometric Shaping with a Single Laser

We have already discussed the use of covariance mapping to determine unambiguously the various fragmentation channels. In I_2, for example, one observes all of the charge-symmetric channels from $I^+ + I^+$ to $I^{5+} + I^{5+}$. There are two aspects of the MEDI process that require explanation. Firstly, the kinetic energies of the atomic ions appears to be *independent* of laser pulse rise time over a large range of pulse lengths – from 0.1 to 2 ps in a recent study [40]. Secondly, for a particular molecule, the dissociation energies of *all* fragments are a specific fraction ($\sim 70\%$ for I_2; 45% for N_2, see earlier) of the Coulomb energy (given by $E = Q_1 Q_2 / R_e$, where Q_1 and Q_2 are the atomic ion charges and R_e is the equilibrium internuclear separation of the neutral molecule). This strongly suggests that all Coulomb explosions occur at a molecule-specific critical distance, R_c, larger than R_e ($R_c = R_e/70\%$ for I_2).

The Saclay group has pointed out that laser-induced stabilisation can provide an explanation for the above facts [41]. The approach by the Ottawa [42] and Reading [43] groups, on the other hand, does not require stabilisation; inertial confinement and extremely rapid sequential ionization at, or near, the critical distance provides the basis of their explanation. Since both the stabilisation model and inertial confinement model invoke multiple ionization close to the critical distance, they cannot easily be distinguished by measuring the kinetic energies of the ions. However, by manipulating the laser pulse rise time interferometrically, it is possible to observe changes in kinetic energies [44].

Varying pulse rise time by simply changing the pulse duration is not straightfoward because the laser pulses may differ in other respects. The principle of the interferometric method is to split each laser pulse into two pulses of unequal intensity, delay the main pulse and recombine them such that the low-intensity prepulse can interfere constructively or destructively with the leading edge of the main pulse, producing an increase or decrease in rise time.

The advantage of the method is that it is external to the laser system and allows rapid and controlled adjustment of the rise time [45].

4.2 Two Colour Excitation

The irradiation of atoms and molecules with two lasers of different frequency and known relative phase has been shown not only to throw light on the ATI process in atoms [46] but has produced a dramatic enhancement in the conversion efficiency in high harmonic generation [47]. In addition, there is theoretical evidence that molecular dynamics may be controlled using two-colour excitation [48].

In the tunnelling regime, the total ionization might be expected to depend solely on the instantaneous electric field. However, the Corkum model [39] suggests that when two laser frequencies are applied to an atom, certain aspects of the ionization process depend, in addition, on the relative phases of the two fields. Using a mixture of fundamental and second harmonic of the Nd:YAG laser, there is recent experimental evidence that the ATI spectra of Kr depend upon the relative phases of the fields [46]. In particular there is a considerable forward-backward asymmetry in the emission of the higher energy electrons.

Of interest in the context of this paper is the possibility of controlling dissociation pathways via control of the phase difference between two laser fields. There has been recent theoretical evidence in H_2^+ that the dissociation probability and the energy and angular distributions of emitted protons are very sensitive to the relative phases of the fundamental and its harmonic [48]. In an experiment on the dissociation of HD^+ in a laser field produced by the superposition of the fundamental and second harmonic of a 50ps, mode-locked, Nd:YLF laser, the charged fragment was ejected counter-intuitively, i.e. in a direction opposite to the maximum in the laser E-field [49]. This behaviour has been explained in terms of electron localisation at the critical internuclear distance [50].

Another experiment one can contemplate involves the manipulation of the fragmentation mode of highly ionized molecules. We have mentioned earlier the discussions with regard to the symmetry of fragmentation of simple diatomic molecules. With the fundamental alone, covariance maps of I_2 should indicate the usual predominance of charge-symmetric fragmentation. With the second harmonic superimposed to produce a forward-backward asymmetry in the resultant E field, the tunnel ionization process should be asymmetric, relative to the detector. That is, the electrons should be ejected preferentially along the direction of highest E field. Since there is no time for rotation of the fragment ion pairs, the ions themselves should be emitted with a forward-backward asymmetry. This behaviour will be reflected in the relevant features on the covariance map.

5 High Harmonic Generation

The detailed study of high harmonic generation has two main aims, the understanding of the process itself and the application of the high harmonics as a coherent, short-pulse, tunable source at vacuum ultraviolet and soft X-ray energies. The generation of energetic photons by frequency conversion in rare-gas media has now been demonstrated in several laboratories [51 and references therein] and extremely high harmonics have been observed. For example, the 135th harmonic of the Nd:glass laser [52] produces radiation at about 8 nm, whilst the 109th harmonic of the Ti:sapphire laser [53] produces radiation at about 7 nm.

A typical experimental arrangement for observing high harmonics is shown schematically in Fig. 10. In this case radiation from a pulsed Nd:glass laser is weakly focused ($f = 1.7$ m) just below the nozzle of a pulsed jet. The gas pressure in the beam is typically 2 torr and the harmonics are studied with a monochromator incorporating a concave grating designed to produce an aberration-corrected, flat focal field [54]. The detector is a pair of microchannel plates (MCP) plus phosphor followed by a CCD camera. Using this arrangement, the angular distribution of the various harmonics can be obtained. A harmonic spectrum using He gas is shown in Fig. 11. Although the spectrum is uncorrected for monochromator/detector efficiency, one observes the typical plateau of harmonics extending to high values of q (the harmonic number), with a rather sharp cut-off at about $q = 115$.

By performing time-dependent calculations for a number of atomic systems, Krause et al. [55] predicted that the width of the plateau should vary as $I_p + 3U_p$, where I_p is the ionization potential of the atom and $U_p = e^2E^2/4m\omega^2$ is the ponderomotive energy, that is the cycle-averaged oscillation energy of a free electron (charge e, mass m) in the laser field E, at a laser frequency ω. The reason for such a formula has been given by Corkum [39] using a quasi-classical

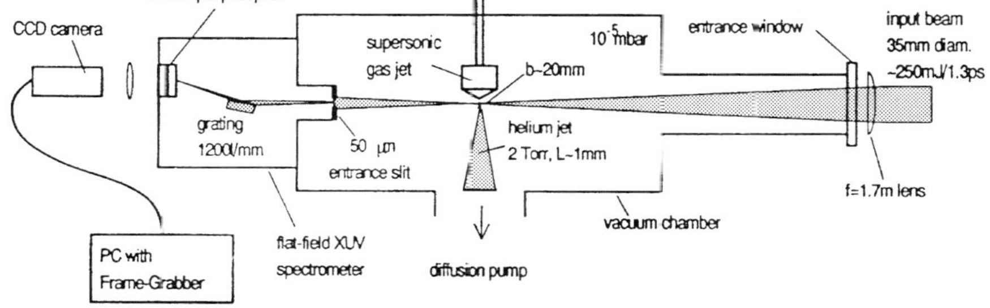

Fig. 10. Schematic diagram of experimental layout for detection of high harmonics

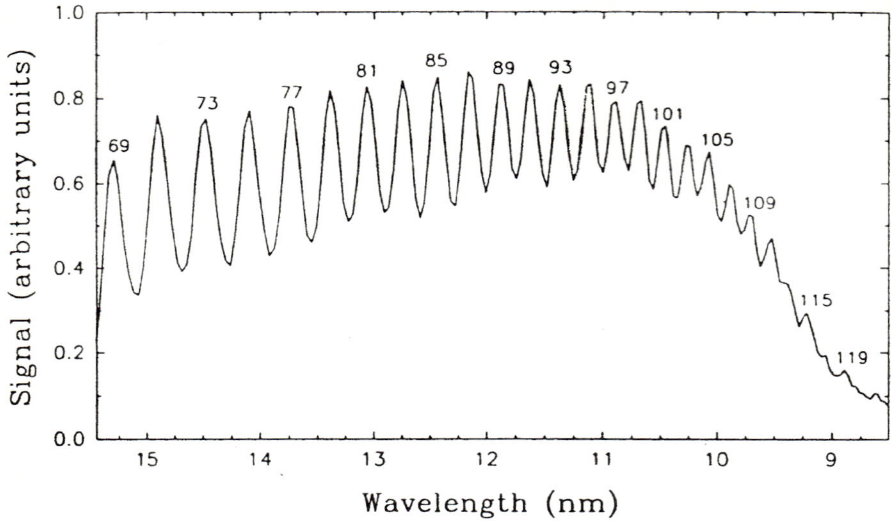

Fig. 11. Observed spectrum of high harmonics in He

approach. The electron is first imagined to tunnel through the barrier formed by the atomic potential and laser field but then the subsequent 'wiggle' motion is treated classically. Thus, in the high intensity, low frequency regime the process of high harmonic generation can be described in terms of a three-step process: tunnel ionization, acceleration of the electron in the laser field, and scattering by the atomic potential leading to recombination to the ground state of the atom.

It is clear from the above formula that the highest harmonics will be observed with atoms of high ionization potential, such as He. Unfortunately, the intensities of these higher harmonics are considerably lower than the plateau harmonics of Ne and Xe. One might also contemplate using ionized atoms. However, the intensity of the harmonics from ions are considerably lower due to the reduced efficiency of the single-ion response and to the poorer phase matching in ionized media, influenced by the free electron dispersion.

For the rare gases, the conversion efficiency is strongly correlated with the polarisability of the atom and theory suggests that this should be the case for molecules [5]. If this were to be the case, one should be able to obtain high efficiencies by suitable choice of molecular species. However, recent work [56] using a 150fs Ti:sapphire laser at 800 nm indicates that the efficiencies of harmonic generation of gases such as N_2, N_2O and CH_4 are no greater than Ar or Xe, even though their static polarisabilities are much higher.

6 Conclusions

This short paper gives a brief overview of some experimental aspects relating to the dynamics of molecules in intense laser fields (other experimental approaches are discussed in more detail elsewhere in this volume) When simple molecules are subjected to intense laser fields, they undergo MEDI, a process that is little understood. At present there are two quite different models that attempt to explain the fact that the kinetic energies of the fragment ions are almost independent of laser pulse rise time, namely inertial confinement and laser-induced stabilisation. Although it seems clear that bond-hardening, a form of laser-induced stabilisation, may be relevant to H_2, it is not clear whether such a process could apply to highly ionized molecular ions.

Evidently more definitive experiments are required to explain the MEDI process. These will involve the use of correlation techniques, such as the covariance approach outlined here. Additional information will be provided by the use of area detectors to define the ion trajectories [57, 58]. In the future, experiments will involve the coherent control of the dissociation dynamics by using interferometric techniques.

References

1. Maine P, Strickland D, Bado P, Pessot M, Mourou G (1988) IEEE J Quantum Electron 24: 398
2. Freeman RR, Bucksbaum PH (1991) J Phys B 24: 325
3. McPherson A, Gibson G, Jara H, Johann U, Luk TS, McIntyre IA, Boyer K, Rhodes CK (1987) J Opt Soc Am B 4: 595
4. Zavriyev A, Bucksbaum PH, Muller HG, Schumacher DW (1990) Phys Rev A 42: 5500
5. Liang Y, Augst S, Chin SL, Beaudoin Y, Chaker M (1994) J Phys B 27: 5119
6. Giusti-Suzor A, He X, Atabek O, Mies FH (1990) Phys Rev Lett 64: 515
7. Giusti-Suzor A, Mies FH (1992) Phys Rev Lett 68: 3869
8. Frasinski LJ, Codling K, Hatherly PA, Barr J, Ross IN, Toner WT (1987) Phys Rev Lett 58: 2424
9. Mainfray G, Manus C (1991) Rep Prog Phys 54: 1333
10. Brummer P, Shapiro M (1993) "Molecules in Laser Fields" Editor: AD Bandrauk (Dekker; New York)
11. Langley AJ, Noad WJ, Ross IN, Shaikh W (1994) App Opt 33: 3875
12. Perry MD, Landen OL, Szöke A, Campbell EM (1988) Phys Rev A 37: 747
13. Kruit P, Read FH (1983) J Phys E 16: 313
14. Wiley WC, McLaren IH (1955) Rev Sci Instrum 26: 1150
15. Frasinski LJ, Stankiewicz M, Randall KJ, Hatherly PA, Codling K (1986) J Phys B 19: L819
16. Frasinski LJ, Codling K and Hatherly PA (1989) Science 246: 1029
17. Cornaggia C, Normand D, Morellec J (1992) J Phys B 25: L415
18. Laberge M, Dietrich P, Corkum PB (1993) "Ultrafast Phenomena", Vol VIII, Editors: J-L Martin, A Migus, GA Mourou, AH Zewail (Berlin: Springer)
19. Hatherly PA, Stankiewicz M, Frasinski LJ, Cross GM (1994) J Phys B 27: 2993
20. Normand D private communication
21. Lavancier J, Normand D, Cornaggia C, Morellec J, Liu HX (1991) Phys Rev A 43: 1461

22. Codling K, Cornaggia C, Frasinski LJ, Hatherly PA, Morellec J, Normand D (1991) J Phys B 24: L593
23. Hill WT, Zhu J, Hatten DL, Cui Y, Goldhar J, Yang S (1992) Phys Rev Lett 69: 2646
24. Dietrich P, Strickland DT, Corkum PB (1993) J Phys B 26: 2323
25. Hatherly PA, Frasinski LJ, Codling K, Langley AJ, Shaikh W (1990) J Phys B 23: L291
26. Codling K, Frasinski LJ, Hatherly PA (1989) J Phys B 22: L321
27. Normand D, Lompré LA, Cornaggia C (1992) J Phys B 25: L497
28. Dietrich P, Strickland DT, Laberge M, Corkum PB (1993) Phys Rev 47: 2305
29. Frasinski LJ, Hatherly PA, Codling K (1991) Phys Lett A 156: 227
30. Allendorf SW, Szöke A (1991) Phys Rev A 44: 518
31. Bucksbaum PH, Zavriyev A, Muller HG, Schumacher DW (1990) Phys Rev Lett 64: 1883
32. Zavriyev A, Bucksbaum PH, Squier J, Saline F (1993) Phys Rev Lett 70: 1077
33. Frasinski LJ, Stankiewicz M, Hatherly PA, Cross GM, Codling K, Langley AJ, Shaikh W (1992) Phys Rev A 46: R6789
34. Luk TS, Pummer H, Boyer K, Shahidi M, Egger H, Rhodes CK (1983) Phys Rev Lett 51: 110
35. Lambropoulos P (1985) Phys Rev Lett 55: 2141
36. L'Huillier A, Lompré LA, Mainfray G, Manus C (1983) Phys Rev A 27: 2503
37. Walker B, Mevel PE, Yang B, Breger P, Chamberet JP, Antonetti A, DiMauro LF, Agostini P (1993) Phys Rev A 48: R894
38. Charalambidis D, Lampropoulos P, Schröder H, Faucher O, Xu H, Wagner M, Fotakis C (1994) Phys Rev A 50: R2822
39. Corkum PB (1993) Phys Rev Lett 71: 1994
40. Cornaggia C, Lavancier J, Normand D, Morellec J, Agostini P, Chambaret JP, Antonetti A (1991) Phys Rev A 44: 4499
41. Schmidt M, Normand D, Cornaggia C (1994) Phys Rev A 50: 5037
42. Seideman T, Ivanov MY, Corkum PB (1995) Phys Rev Lett 75: 2819
43. Posthumus JH, Frasinski LJ, Giles AJ, Codling K (1995) J Phys B 28: L349
44. Posthumus JH, Giles AJ, Thompson MR, Shaikh W, Langley AJ, Frasinski LJ, Codling K (1996) J Phys B (in press)
45. Giles AJ, Posthumus JH, Thompson MR, Frasinski LJ, Codling K, Langley AJ, Shaikh W, Taday P (1994) Opt Comm 118: 537
46. Schumacher DW, Weihe F, Muller HG, Bucksbaum PH (1994) Phys Rev Lett 73: 1344
47. Zuo T, Bandrauk AD, Ivanov M, Corkum PB (1995) Phys Rev A 51: 3991
48. Charron E, Giusti-Suzor A, Mies FH (1994) Phys Rev A 49: R641
49. Sheehy B, Walker B, DiMauro LF (1995) Phys Rev Lett 74: 4799
50. Posthumus JH, Thompson MR, Giles AJ, Codling K (1996) Phys Rev (in press)
51. Wahlström CG, Larsson J, Persson A, Starczewski T, Svanberg S, Salieves P, Balcou Ph, L'Huillier A (1993) Phys Rev A 48: 4709
52. L'Huillier A, Balcou Ph (1993) Phys Rev Lett 70: 774
53. Macklin JJ, Kmetec JD, Gordon III CL (1993) Phys Rev Lett 70: 766
54. Tisch JGW, Muffett JE, Smith RA, Ciarrocca M, Marangos JP, Hutchinson MHR, Wahlström CG (1994) "Multiphoton Processes" Editors: DK Evans and SL Chin (World Scientific, Singapore)
55. Krause JL, Schafer KJ, Kulander KC (1992) Phys Rev Lett 68: 3535
56. Lynga C, L'Huillier A, Wahlström C-G (1996) ICOMPVII Garmisch-Partenkirchen (Abstract)
57. Werner U, Beckord K, Becker J,. Lutz HO (1995) Phys Rev Lett 74: 1962

Two Interacting Charged Particles in Strong Static Fields: A Variety of Two-Body Phenomena

Peter Schmelcher and Lorenz S. Cederbaum

Theoretische Chemie, Physikalisch-Chemisches Institut, Universität Heidelberg,
Im Neuenheimer Feld 253, 69120 Heidelberg, FRG

In the presence of a homogeneous magnetic field the center of mass and internal motion of a neutral or charged two-body system cannot be separated. We review the effects and phenomena which occur due to the inherent two-body character of these systems. The most prominent effects that occur for neutral species are the chaotic classical diffusion of the center of mass, the intermittent near-threshold behavior and the appearance of weakly bound states with an extraordinarily large electric dipole moment in crossed electric and magnetic fields. For ionic two-body systems the regions of strong mixing of the center of mass and internal motion are identified and investigated. As major effects we observe the classical self-stabilization as well as selfionization processes of the ion.

1 Introduction 28

2 Neutral Two-Body Systems in External Fields 30
2.1 Fundamental Aspects 30
2.2 The Anisotropic Quantum Oscillator in a Magnetic Field 33
2.3 The Hydrogen Atom in Crossed Electric and Magnetic Fields 37
2.4 Classical Diffusion of the Center of Mass 42
2.5 Intermittent Dynamics: a Typical Threshold Phenomenon for Finite Pseudomomentum 46

3 Charged Two-Body Systems in a Magnetic Field 49
3.1 Fundamental Properties 49
3.2 Quantum Mechanical Aspects of the Interaction of the Ionic Center of Mass and Electronic Motions 52
3.3 The classical Self-Stabilization and Self-Ionization Processes 56

4 Summary and Conclusions 60

5 References 62

1 Introduction

The behavior and the properties of interacting particle systems in strong external fields has become a very active field of research in the past two decades. Prominent effects in different areas of physics, for example the quantum Hall effect for a two-dimensional extended electron system or the interplay of regularity and chaos for Rydberg atoms in external fields, have their origin in the action of a strong magnetic field. One of the simplest physical systems, which already reveals much of the complexity of the dynamics in the presence of a magnetic field, is a nonrelativistic interacting two-body system of charged particles subjected to a homogeneous external field. In atomic physics this corresponds either to the neutral hydrogen atom, or to any one electron ion with net charge $Q = (1 - Z)e$ where Z is the nuclear charge number and e is the electron charge. An external field is then characterized as strong if the corresponding magnetic energies/forces are comparable to, or even larger than, the Coulomb binding energies/forces of the underlying system. For the ground states of atoms the typical field strengths are of the order of magnitude or greater than 1 a.u. $= 2.35 \times 10^5$ T, which is an astrophysical field strength and, therefore, only available in the atmosphere of white dwarfs and/or neutron stars. However, with increasing degree of excitation of the atom the level spacing decreases rapidly, and for sufficiently high excited Rydberg states the magnetic interaction energies already dominate over the Coulomb binding energies for laboratory field strengths of a few Tesla. Since highly excited states of the atom correspond to values of the action which are much larger than the elementary quantum of action $\hbar$, a semiclassical or even classical approach to the properties and dynamics of the system becomes reasonable, and can yield important insights into the behavior of excited atoms in strong fields.

During the past twenty years an enormous increase in our knowledge on the hydrogen atom in a strong magnetic field has taken place. In the late 1970s and early 1980s the ground and the first few low-lying excited states of the hydrogen atom were investigated for the whole range of astrophysical field strengths, 10^7–10^{13} Tesla. Subsequently the computational techniques and methods for the calculation of the electronic spectra and wave functions of the atom became more and more elaborate and effective. As a consequence it was possible in the second half of the 1980s to investigate many thousand excited states, and in particular many Rydberg states for which a laboratory magnetic field strength already represents a strong field (see [1] and references therein). Recent developments even allow for a detailed investigation of the electronic states beyond the field-free ionization threshold [2].

All of the above-mentioned investigations on the hydrogen atom were performed in the fixed-nuclei approximation, i.e. for the assumption of an infinitely heavy nucleus. In the absence of a magnetic field the center of mass (CM) and electronic motion separate exactly, and the influence of the finite nuclear mass on the behavior and properties of the atom can be taken into account simply by replacing the electron mass by the corresponding reduced mass. In the presence of a magnetic field, however, such a separation of the CM and internal motion is not possible, i.e. the CM and electronic motion are intimately coupled. The latter coupling reflects the inherent two-body character of a system of two interacting charged particles in an external magnetic field. Since this coupling is proportional to the inverse of the total mass of the system, the corresponding term disappears for the artificial assumption of an infinitely heavy nucleus. Within the present article we review different two-body phenomena which have been discovered and studied for neutral as well as charged two-body systems in strong static fields. This refers to investigations which have been performed during the past five years. Due to the coupling of the CM and electronic degrees of freedom or, more generally, due to the finite nuclear mass, it will become evident that the inclusion of the dynamics of the heavy nucleus is crucial in the presence of a magnetic field, and yields a variety of interesting and surprising phenomena.

The article is organized as follows. Since the two cases of a neutral and charged system are fundamentally different, we divide our article into two main parts: Sect. 2 deals with neutral two-body systems, whereas the case of a charged system will be discussed in Sect. 3. Section 2 takes on the following structure. In Sect. 2.1 we give an introduction to the analytical foundations and establish in particular the existence of an outer potential well for neutral two-body systems in crossed electric and magnetic fields. In Sect. 2.1 the problem of the anisotropic quantum oscillator in a magnetic field is discussed and solved analytically in closed form. This serves as a most helpful starting point for the *quantum mechanical* investigations in Sect. 2.3 which deal with phenomena related to the outer potential well of the hydrogen atom in crossed electric and magnetic fields. Subsequently, we investigate in Sect. 2.4 the changes in the behavior of the *classical* CM motion which are induced by the transition from regularity to chaos in the internal motion of the atom. In this section we exclusively discuss the case of vanishing pseudomomentum. Section 2.5 finally contains an elaborate discussion of the intermittent classical dynamics which occurs for the hydrogen atom with a sufficiently large pseudomomentum and/or an additional perpendicular electric field. Section 3 again starts with a discussion of the analytical foundations which in particular contain a transformation of the ionic Hamiltonian to an elucidating structure. This Hamiltonian will be used throughout the whole section in order to investigate the interaction of the ionic CM and the electronic motion. In Sect. 3.2 we will establish those regions in parameter space (energy, field strength, etc ...) where a strong *quantum mechanical* mixing (or coupling) of the CM and electronic wave functions occurs. Section 3.3 is devoted to a study of the *classical dynamics* of the ion. Depending on the CM and electronic energies as well as the field strength, we observe in particular a self-stabilization effect as well as the process of the self-ionization of the ion via energy transfer from the CM to the electronic degrees of freedom. The final section, Sect. 4 presents the conclusions.

2 Neutral Two-Body Systems in External Fields

2.1 Fundamental Aspects

The general problem we are concerned with is a neutral system of two oppositely charged particles interacting via the Coulomb potential in the presence of an external homogeneous magnetic field and an additional electric field oriented perpendicular to the magnetic one. The corresponding Hamiltonian takes on the following appearance:

$$\mathscr{H}_L = \sum_{i=1}^{2} \left[\frac{1}{2m_i} (\mathbf{p}_i - e_i \mathbf{A}_i)^2 - e_i \mathbf{E} \mathbf{r}_i \right] + V(|\mathbf{r}_1 - \mathbf{r}_2|) \tag{1}$$

where e_i, m_i, $\mathbf{A}_i$ and $\mathbf{E}$ denote the charges, masses, vector potential and electric field vector, respectively. $\{\mathbf{r}_i, \mathbf{p}_i\}$ are the Cartesian coordinates and momenta in the laboratory coordinate system. In the absence of the external fields the total canonical momentum is conserved and equals the total kinetic momentum. The resulting symmetry group is formed by the translations in coordinate space. In CM and internal variables the CM motion decouples, i.e. separates completely from the relative motion.

In the presence of the external fields the vector potential appears in the Hamiltonian at Eq. (1) and the space translation symmetry is, therefore, lost. However, there exists a generalization, i.e. the phase space translation group [3] which provides a symmetry associated with the CM motion of the system in the presence of the external fields. The new conserved quantity which is the corresponding generalization of the total canonical momentum of the field-free case is the so-called pseudomomentum $\mathbf{K}$ [3, 4]

$$\mathbf{K} = \sum_{i=1}^{2} (\mathbf{p}_i - e_i \mathbf{A}_i + e_i \mathbf{B} \times \mathbf{r}_i) - M\mathbf{v}_D \tag{2}$$

where M and $\mathbf{B}$ are the total mass and magnetic field vector, respectively and the term $M\mathbf{v}_D$ has been included in the pseudomomentum for convenience. $\mathbf{v}_D = (\mathbf{E} \times \mathbf{B})/B^2$ is the drift velocity of free charged particles in crossed fields. The latter is independent of the charge and mass of the particles [5]. Since the pseudomomentum is conserved its components commute with the Hamiltonian at Eq. (1)

$$[\mathbf{K}_\alpha, \mathscr{H}_L] = 0 \tag{3}$$

and for a neutral system $e_1 = -e_2 = e$, i.e. vanishing total charge, the components of $\mathbf{K}$ also commute among themselves:

$$[\mathbf{K}_\alpha, \mathbf{K}_\beta] = 0\,. \tag{4}$$

As a consequence the components of $\mathbf{K}$ can be made sharp simultaneously. This fact is the key ingredient for the so-called pseudoseparation of the CM motion which introduces the pseudomomentum as a canonical momentum and thereby eliminates the canonical CM coordinate. However, this does not mean that the CM and internal degrees of freedom decouple, i.e. are separated. The pseudoseparation merely uses the above exact constants of motion of the Hamiltonian in order to transform the coupling of the collective and internal degrees of freedom to a particularly simple form. The pseudoseparation transformation consists of a combined coordinate and unitary gauge transformation. In the literature [3, 4, 6] this transformation has been performed with the assumption of a fixed gauge for the vector potential in the Hamiltonian. However, fixing the gauge from the very beginning, i.e. already for the Hamiltonian in the laboratory coordinate system, possesses a serious drawback. It is not possible to discern between gauge dependent and gauge invariant terms in the transformed Hamiltonian. A gauge invariant pseudoseparation is, therefore, desirable and has been performed very recently [7]. As a result of this gauge independent pseudoseparation we obtain the transformed Hamiltonian

$$\mathscr{H} = \mathscr{T} + \mathscr{V} \tag{5}$$

where

$$\mathscr{T} = \frac{1}{2\mu}(\mathbf{p} - q\mathbf{A}(\mathbf{r}))^2 \tag{6}$$

and

$$\mathscr{V} = \frac{1}{2M}(\mathbf{K} - e\mathbf{B}\times\mathbf{r})^2 + V(r) + \frac{M}{2}\mathbf{v}_D^2 + \mathbf{K}\mathbf{v}_D \tag{7}$$

with the charge $q = \frac{e\mu}{\hat{\mu}}$ where $\mu = \frac{mM_0}{M}$ and $\hat{\mu} = \frac{mM_0}{M_0 - m}$ are different reduced masses. $\mathbf{K}$ is now the constant vector of the pseudomomentum and $\{\mathbf{r}, \mathbf{p}\}$ denote the canonical pair of variables for the internal relative motion The Hamiltonian (5) is the sum of two terms: the kinetic energy $\mathscr{T}$ of the relative motion and the potential $\mathscr{V}$.

The explicit form of the kinetic energy $\mathscr{T}$ depends on the chosen gauge via the vector potential $\mathbf{A}$. The important novelty with respect to our Hamiltonian $\mathscr{H}$ is the appearance of the potential term $\mathscr{V}$. Apart from the Coulomb potential V and the constant term $\frac{M}{2}\mathbf{v}_D^2 + \mathbf{K}\mathbf{v}_D$ there occurs in the total potential $\mathscr{V}$ an additional potential term

$$V_0 = \frac{1}{2M}(\mathbf{K} - e\mathbf{B}\times\mathbf{r})^2 . \tag{8}$$

The latter term is gauge independent, i.e. does not contain the vector potential, and, therefore, fully deserves the interpretation of an additional potential term for the internal motion with the kinetic energy at Eq. (6). Apart from the constant $\frac{\mathbf{K}^2}{2M}$ the potential V_o contains two coordinate dependent parts. The term linear in the coordinates $-\frac{e}{M}(\mathbf{K}\times\mathbf{B})\mathbf{r}$ consists of two Stark terms: one which is due to the external electric field $\mathbf{E}$ and a second one which is a motional Stark

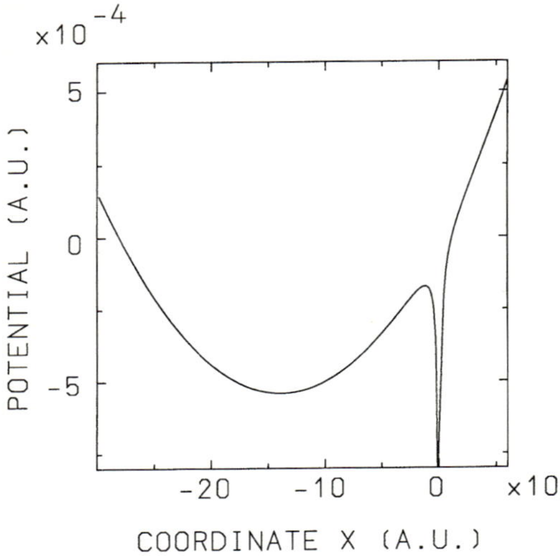

Fig. 1. The potential $\mathcal{V}$ (see Eq. (7)) along the direction of the motional electric field. Values of the pseudomomentum and magnetic field strength are $\mathbf{K} = (0, 1.4, 0)$ and $\mathbf{B} = (0, 0, 10^{-5})$ in atomic units ($B = 1$a.u. corresponds to 2.35×10^5 T)

term with an induced constant electric field $\frac{1}{M}((\mathbf{K} + M\mathbf{v}_D) \times \mathbf{B})$. The latter electric field is oriented perpendicular to the magnetic one and arises due to the collective motion of the atom through the homogeneous magnetic field. Besides the linear terms there exists a quadratic, i.e. diamagnetic, term $\frac{e^2}{2M}(\mathbf{B} \times \mathbf{r})^2$ in the potential V_o.

In the following we illustrate and discuss the qualitative properties of the potential $\mathcal{V}$. Figure 1 shows for the choice $\mathbf{B} = (0, 0, B)$, $\mathbf{K} = (0, K, 0)$ and a vanishing external electric field an intersection of the potential $\mathcal{V}$ along the direction of the motional electric field ($y = z = 0$). For small values of the x-coordinate the term $(\frac{1}{|x|})$ due to the Coulomb potential dominates. With increasing values of the coordinate x the Stark term $(\frac{-e}{M}) BKx$ increases and becomes comparable with the strength of the Coulomb potential. The diamagnetic term $(\frac{e^2}{2M}) B^2 x^2$ is, for our choice of the parameter values (see Sect. 2.3), still negligible in this coordinate region. Due to the competition of the Coulomb and Stark term a saddle point arises which is located at approximately $x \approx -1.1 \times 10^4$ a.u. in Fig. 1. For even larger absolute values of the x coordinate the Coulomb potential becomes small and the shape of the potential $\mathcal{V}$ is more and more determined by the diamagnetic quadratic potential term $(\frac{e^2}{2M}) B^2 x^2$. Due to the competition of the Stark and diamagnetic terms our potential $\mathcal{V}$ now develops a minimum which is located in Fig. 1 at approximately $x \approx -1.4 \times 10^5$ a.u. The existence of both the saddle point and the minimum depends, of course, on the values of the magnetic field strength and the pseudomomentum. For a derivation and discussion of the conditions for their existence we refer the reader to Sect. 2.3 and to the literature [7–10].

The above-mentioned properties of our potential $\mathcal{V}$ have important implications on the dynamical behavior of the atom. First of all, we observe that the ionization of the atom can take place only in the direction of the magnetic field axis. In the direction perpendicular to the magnetic field vector the diamagnetic term $\frac{e^2}{2M}(\mathbf{B} \times \mathbf{r})^2$ is dominating for large distances $\rho = (x^2 + y^2)^{1/2}$ and causes a confining behavior of the potential $\mathscr{V}$. Therefore, ionization is not allowed perpendicular to the magnetic field. The second important observation is the fact that the presence of a minimum in the potential $\mathscr{V}$ leads to a potential well and consequently to new bound states or trajectories inside this well. These trajectories and the corresponding quantum mechanical states are extended objects in the sense that the electron and the nucleus are far from each other. The quantum dynamics inside this well will be discussed in Sect. 2.3 whereas the classical dynamics for energies beyond the saddle point energy is the subject of investigation in Sect. 2.5.

Finally, we emphasize that the above potential V_o is inseparably connected with the finite nuclear mass. Assuming an infinite nuclear mass would simply yield $V_o \equiv 0$ and the above-mentioned properties of the total potential $\mathscr{V}$ would disappear. In order to obtain the correct qualitative behavior and properties of the atom in a strong magnetic field it is, therefore, necessary to treat the atom as an inseparable two-body system.

2.2 The Anisotropic Quantum Oscillator in a Magnetic Field

In the present section we consider a particle with charge q in a static homogeneous magnetic field $\mathbf{B} = B\mathbf{e}_z$ and an anisotropic harmonic potential (see in particular [7]). We will provide closed-form analytical solutions of the corresponding Schrödinger equation, i.e. explicit expressions for the energy eigenvalues and eigenfunctions. The results on the oscillator system will be most helpful for our discussion of the dynamics in the outer potential well of the hydrogen atom in crossed electric and magnetic fields in Sect. 2.3. In particular the Hamiltonian of two oppositely charged particles in a magnetic field interacting with an anisotropic harmonic potential can, according to the pseudoseparation discussed in Sect. 2.1, be transformed to the anisotropic quantum oscillator in a magnetic field (see Eq. (5)).

We will assume in the following that one of the principal axes of the anisotropic harmonic potential points along the direction of the magnetic field. For the vector potential we choose the symmetric gauge $\mathbf{A} = \frac{1}{2}\mathbf{B} \times \mathbf{r}$. The Hamiltonian of the system is given by

$$
\begin{aligned}
H = {} & \frac{\mathbf{p}^2}{2\mu} - \frac{qB}{2\mu}(xp_y - yp_x) + \frac{q^2B^2}{8\mu}(x^2 + y^2) \\
& + \frac{\mu}{2}(\omega_x^2 x^2 + \omega_y^2 y^2 + \omega_z^2 z^2). \qquad (9)
\end{aligned}
$$

Obviously the z-direction can be separated from the x and y directions, i.e. the Hamiltonian H is a sum of the Hamiltonian H_{xy} of the two-dimensional system perpendicular to the magnetic field and the Hamiltonian H_z of the one-dimensional harmonic oscillator in the z-direction. The eigenvalues of H are given by

$$E_{n_1 n_2 n_z} = E_{n_1 n_2} + E_{n_z} \tag{10}$$

where $E_{n_z} = (n_z + \frac{1}{2})\omega_z$ and $E_{n_1 n_2}$ are the eigenvalues of H_{xy}. Accordingly, the eigenfunctions of H are given by the product of the eigenfunctions of the one-dimensional harmonic oscillator, $\phi_{n_z}(z)$, and the eigenfunctions $\mathbf{\Psi}_{n_1 n_2}(x, y)$ of H_{xy}. Both the eigenvalues and eigenfunctions of H_{xy} will be derived in the following.

The existence of the angular momentum term $\frac{qB}{2\mu}(xp_y - yp_x)$ in H_{xy} makes the solution of the corresponding Schrödinger equation a non-trivial task. Because of the loss of symmetry caused by the anisotropic harmonic potential, the angular momentum is not a conserved quantity. Therefore, its quantum number cannot be used to reduce the two-dimensional problem to a one-dimensional differential equation as in the case of an isotropic potential [11]. We have to find a unitary transformation of the Hamiltonian that eliminates the angular momentum term and the explicit dependence on the magnetic field in H_{xy}. By doing this we will end up with a Hamiltonian which has the usual form of the kinetic energy in the absence of a magnetic field, i.e. $\frac{p^2}{2m}$, and potential terms that do not contain momentum operators. In fact, the final Hamiltonian will describe two independent particles in their individual one-dimensional harmonic potentials. The corresponding eigenvalues and functions are well known.

The time-independent Schrödinger equation for the Hamiltonian H_{xy}

$$H_{xy} \mathbf{\Psi}_{n_1 n_2}(x, y) = E_{n_1 n_2} \mathbf{\Psi}_{n_1 n_2}(x, y) \tag{11}$$

can be written in the form

$$H_3 \psi_{n_1 n_2} = E_{n_1 n_2} \psi_{n_1 n_2} \tag{12}$$

where we have introduced the eigenfunctions $\psi_{n_1 n_2}$ of the Hamiltonian $H_3 = U^{-1} H_{xy} U$ defined by

$$\mathbf{\Psi}_{n_1 n_2}(x, y) = (U \psi_{n_1 n_2})(x, y) \tag{13}$$

with the unitary operator

$$U = \exp(i\alpha xy) \exp(i\beta p_x p_y) . \tag{14}$$

H_3 can be obtained by using the transformation formulae given in [12]. We want the transformed Hamiltonian H_3 to describe a system of two independent particles in one-dimensional harmonic potentials. This fixes the values of α and β. We obtain after some tedious algebra explicit expressions for the coefficients α and β and in particular the explicit form of H_3

$$H_3 = \frac{p_x^2}{2M_1} + \frac{p_y^2}{2M_2} + \frac{M_1}{2} \omega_1^2 x^2 + \frac{M_2}{2} \omega_2^2 y^2 \tag{15}$$

with the masses

$$M_{1,2} = \frac{2\mu\sqrt{(\omega_x^2 + \omega_y^2 + \omega_c^2)^2 - 4\omega_x^2\omega_y^2}}{\text{sgn}[\omega_x^2 - \omega_y^2](\omega_x^2 - \omega_y^2 \pm \omega_c^2) + \sqrt{(\omega_x^2 + \omega_y^2 + \omega_c^2)^2 - 4\omega_x^2\omega_y^2}} \tag{16}$$

where

$$\text{sgn}[\omega_x^2 - \omega_y^2] = \begin{cases} +1 & \text{for } \omega_x^2 \geqslant \omega_y^2 \\ -1 & \text{otherwise.} \end{cases} \tag{17}$$

For the frequencies ω_1 and ω_2 we obtain the expression

$$\omega_{1,2} = \frac{1}{\sqrt{2}}\sqrt{\omega_x^2 + \omega_y^2 + \omega_c^2 \pm \text{sgn}[\omega_x^2 - \omega_y^2]\sqrt{(\omega_x^2 + \omega_y^2 + \omega_c^2)^2 - 4\omega_x^2\omega_y^2}}\,. \tag{18}$$

Now, the original two-dimensional one-particle Hamiltonian H_{xy} which mixes the x- and y-directions via the angular momentum term has been transformed into the sum of two Hamiltonians of independent particles with masses M_1 and M_2 in one-dimensional harmonic potentials with frequencies ω_1 and ω_2, respectively. The new masses and frequencies are given in terms of the original frequencies ω_x, ω_y, and the cyclotron frequency $\omega_c = \frac{-qB}{\mu}$. Via ω_c, there is an implicit dependence of the Hamiltonian H_3 on the magnetic field strength B.

The eigenvalues of H_3, which are also those of H_{xy}, are given by

$$E_{n_1 n_2} = \left(n_1 + \frac{1}{2}\right)\omega_1 + \left(n_2 + \frac{1}{2}\right)\omega_2. \tag{19}$$

The eigenfunctions of H_3 are products

$$\psi_{n_1 n_2}(x, y) = \phi_{n_1}(x)\phi_{n_2}(y) \tag{20}$$

of the one-dimensional eigenfunctions ϕ_{n_i} of the harmonic oscillators with frequencies ω_i and masses M_i.

In order to obtain the eigenfunctions of H_{xy} we have to apply the transformation operator U (Eq. (14)) to $\psi_{n_1 n_2}$ according to Eq. (13). This is a non-trivial task. The operator $\exp(i\beta p_x p_y)$ acts on a function of the coordinates x and y and the result cannot be derived directly. It is very simple, though, to get the result if the operator acts on a function in momentum space. For the determination of the wave function we, therefore, proceed as follows. First we take the Fourier transform of the eigenfunction $\psi_{n_1 n_2}$ of H_3. As a next step we apply the operator $\exp(i\beta p_x p_y)$ which is in momentum space a simple multiplication. Next we take the inverse Fourier transform of the result in order to obtain the function $f_{n_1 n_2}(x, y)$ in the coordinate space of H_{xy}. In the latter step we use the convolution theorem [13] for Fourier transforms. Subsequently applying the product expansion for the Hermite polynomials $\mathscr{P}_i$ with shifted arguments and acting finally with the operator $\exp(i\alpha xy)$ yields the following result (for details of the calculation we refer the reader to [7]):

$$\Psi_{n_1 n_2}(x, y) = N e^{g(x,y)} \sum_{k=0}^{n_1} \sum_{l=0}^{n_2} c_{kl}(n_1, n_2) \mathscr{P}_{n_1-k}(\sqrt{2}F)\, \mathscr{P}_{n_2-l}(\sqrt{2}J) \tag{21}$$

where

$$N = \frac{(-i)^{n_2}}{2^{n_1+n_2}\pi} \sqrt{\frac{\sqrt{M_1\omega_1 M_2\omega_2}}{n_1! n_2! (\beta^2 M_1\omega_1 M_2\omega_2 + 1)}} \tag{22}$$

and

$$g(x, y) = i\alpha xy - \frac{M_1\omega_1 x^2 + 2i\beta M_1\omega_1 M_2\omega_2 xy + M_2\omega_2 y^2}{2(\beta^2 M_1\omega_1 M_2\omega_2 + 1)}. \tag{23}$$

The coefficients $c_{kl}(n_1, n_2)$ take on the appearance

$$c_{kl}(n_1, n_2) =$$

$$\begin{cases} 0 & \text{for } k + l \text{ odd} \\ \left. \begin{aligned} & 2^{2l+k}\binom{n_1}{k}\binom{n_2}{l}\Gamma\left(\frac{k+l+1}{2}\right) D^l G^l (2D^2 - 1)^{\frac{k-l}{2}} \\ & \times {}_2F_1\left(-\frac{l}{2}, \frac{1-l}{2}; \frac{1-k-l}{2}; \frac{1}{2D^2} + \frac{1}{2G^2} - \frac{1}{4D^2G^2}\right) \end{aligned} \right\} & \text{otherwise.} \end{cases}$$

Here ${}_2F_1$ denotes the hypergeometric function. D and G are the constants

$$D = \sqrt{\frac{2\beta^2 M_1\omega_1 M_2\omega_2}{\beta^2 M_1\omega_1 M_2\omega_2 + 1}} \tag{24}$$

$$G = -\operatorname{sgn}[\omega_x^2 - \omega_y^2] \sqrt{\frac{2}{\beta^2 M_1\omega_1 M_2\omega_2 + 1}}. \tag{25}$$

F and J are two complex functions:

$$F = \frac{\sqrt{M_1\omega_1}(x + i\beta M_2\omega_2 y)}{\beta^2 M_1\omega_1 M_2\omega_2 + 1} \tag{26}$$

$$J = \frac{\sqrt{M_2\omega_2}(\beta M_1\omega_1 x - iy)}{\beta^2 M_1\omega_1 M_2\omega_2 + 1}. \tag{27}$$

$\Psi_{n_1 n_2}(x, y)$ are the eigenfunctions of H_{xy} given in coordinate space. They are finite sums of products of two Hermite polynomials with different complex arguments.

For a comparison of the classical trajectories and the above-mentioned quantum mechanical wave functions of the anisotropic oscillator in a magnetic field we refer the reader to the literature [7].

2.3 The Hydrogen Atom in Crossed Electric and Magnetic Fields

Having determined the eigenvalues and eigenfunctions of a charged particle in a magnetic field and an anisotropic harmonic potential we now proceed to consider a real physical system, the hydrogen atom. In Sect. 2.1 we showed that, in addition to the Coulomb singularity, there exists an outer potential well for the relative motion. This well is approximately an anisotropic harmonic potential in the vicinity of its minimum. Accordingly, we can use the results of the preceding section and compare them to the numerically calculated exact eigenstates of the hydrogen atom in the well. Before doing this let us briefly discuss the conditions for the existence of the outer potential well [7–10].

Assuming the same orientations of the magnetic field vector and pseudomomentum as in Fig. 1 we obtain from the condition for a potential extremum $\frac{\partial \mathscr{V}}{\partial r_i} = 0$ (see Eq. (7)) for the y and z coordinate $y_0 = z_0 = 0$ and an equation for the x coordinate

$$x_0^3 + \frac{K}{B} x_0^2 - \frac{M}{B^2} = 0 \tag{28}$$

where $x_0 < 0$. In order to get both a minimum and a saddle point the cubic equation must have three real zeros. From the form of the discriminant we obtain the inequality

$$K^3 > \frac{27}{4} BM \tag{29}$$

as a necessary condition for the existence of a minimum. In the literature [8] an explicit approximation formula has been given for the minimum coordinate: $x_0 \simeq -\frac{K}{B} + \frac{KM}{K^3 - 2MB}$. Hence, for laboratory field strengths ($B \sim 10^{-5}$ a.u.) and pseudomomenta of the order of 1 a.u., the minimum is located at a distance of about 10^5 a.u. from the Coulomb singularity. Therefore, for states in the well the electron and proton are separated about 100 000 times as much as they are in the ground state of the hydrogen atom without external fields, i.e. we encounter a strongly delocalized atom of almost macroscopic size. Since the well exists only for negative values of x the separation is fixed in a certain direction of space resulting in a large permanent dipole moment of the atom in contrast to the well known Rydberg states with vanishing pseudomomentum in a magnetic field that do not exhibit a permanent dipole moment.

In the following we will investigate the quantum mechanical states in the outer potential well [7–10] for laboratory field strengths. Many states up to a very high degree of excitation will be considered [7]. In order to interpret the full numerical results for the energies and wave functions of the hydrogen atom in a magnetic field (or alternatively crossed electric and magnetic fields) let us perform an expansion of the Coulomb potential $\frac{1}{|\mathbf{r}|}$ around the minimum of the potential well. Including only terms up to x_0^{-3} we get the approximated potential [8]

$$V_h = \frac{\mu}{2}\omega_x^2 x^2 + \frac{\mu}{2}\omega_y^2 y^2 + \frac{\mu}{2}\omega_z^2 z^2 + C_3 \tag{30}$$

where we used new coordinates with the origin at the minimum of the well. The frequencies are given by

$$\omega_x = \sqrt{\frac{2}{\mu}\left(\frac{B^2}{2M} + \frac{1}{x_0^3}\right)}, \quad \omega_y = \sqrt{\frac{1}{\mu}\left(\frac{B^2}{M} - \frac{1}{x_0^3}\right)} \tag{31}$$

$$\omega_z = \sqrt{\frac{1}{\mu}\left(-\frac{1}{x_0^3}\right)}$$

and the constant reads $C_3 = \frac{2}{x_0} - \frac{B^2 x_0^2}{2M} + \frac{K^2}{2M} + \frac{ME^2}{2B^2} + \frac{KE}{B}$. With this potential the Hamiltonian of the hydrogen atom takes on the form of the Hamiltonian of a charged particle in a magnetic field with anisotropic harmonic interaction (see Eq. (9)). Therefore, we can use the analytical results of Sect. 2.2 as an approximation for the low-lying energies and wave functions of the hydrogen atom in the well. The approximate energies are given by

$$E_{n_+ n_- n_z} = \left(n_+ + \frac{1}{2}\right)\omega_+ + \left(n_- + \frac{1}{2}\right)\omega_- + \left(n_z + \frac{1}{2}\right)\omega_z + C_3 \tag{32}$$

with the frequencies

$$\omega_{+,-} = \frac{1}{\sqrt{2}}\sqrt{\omega_x^2 + \omega_y^2 + \omega_c^2 \pm \sqrt{(\omega_x^2 + \omega_y^2 + \omega_c^2)^2 - 4\omega_x^2\omega_y^2}}\,. \tag{33}$$

The quantum numbers n_+, n_- and n_z apply to the eigenstates of the Hamiltonian with the harmonic potential V_h only. However, we will use them as labels for the states of the hydrogen atom, too. For all calculated cases ω_+ will be much larger than ω_-, therefore n_+ will always be zero for the hundreds of states considered here. An increase of the value of the label n_- will correspond to an increase of the extension of the wave function in the x- and y-directions, and an increase of the value of the label n_z will correspond to an increase of the extension of the wave function in the z-direction.

In a second step, in order to determine the influence of the anharmonicity in the exact potential we will expand the $\frac{1}{|\mathbf{r}|}$- term up to higher powers of the components of $\mathbf{r}$ and treat them as small perturbations to the harmonic approximation of the Hamiltonian by means of first order perturbation theory. These perturbative calculations offer insight into the effects of the anharmonic parts of the potential onto the energies and the form of the wave functions. For a discussion of the basis set method and the computational techniques used for the numerical calculation of the exact eigenenergies and eigenfunctions in the outer potential well we refer the reader to [7]. In the following we discuss the results of these numerical calculations of the exact eigenenergies and wave functions and compare them to the corresponding quantities obtained by the harmonic approximation as well as the perturbation theoretical calculation.

Our choice of the parameter values is $K = 0.6$, $B = 10^{-5}$ and $E = 0$. The energy of the ground state in the potential well is $E_0 = -1.2499 \times 10^{-5}$ a.u. The binding energy of the ground state is $E_B = E_S - E_0 = 1.75 \times 10^{-5}$ a.u. where E_S is the threshold energy, i.e. the lowest energy of the ionized system. Even though the binding energies of the states in the well are relatively weak they should be stable as long as collisional interaction is prevented. The energy gap between the ground and first excited state corresponds to a frequency of the order of magnitude of a few tens of MHz.

In the following we consider the deviation of the exact energies from those of the harmonic approximation as a function of the level number. In Fig. 2 we have plotted the energy difference between the harmonic approximation and the exact energies of the hydrogen atom in units of ω_-. We see that the difference grows stepwise while neighboring states show very different deviations from the harmonic approximation. To explain these features let us look at the energy level number 162. We see that the difference between exact and approximated energies for this level is much larger than for the levels below 162. Level 162 has

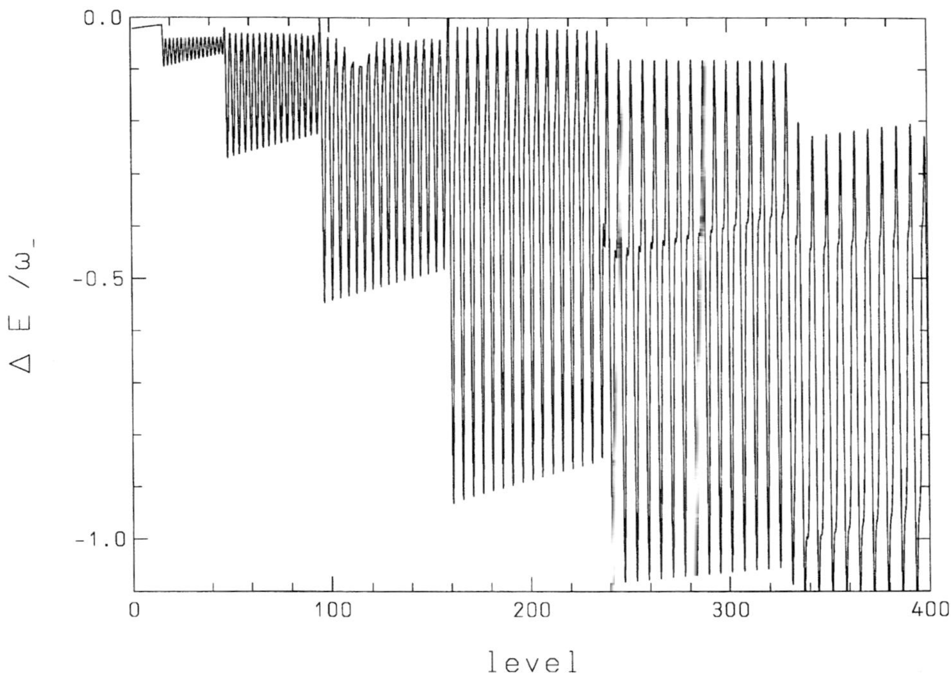

Fig. 2. Difference between harmonic approximated and exact energies of a hydrogen atom in a magnetic field in units of the frequency ω_- as a function of the energy level. Parameters: $B = 10^{-5}$, $K = 0.6$ in atomic units

the quantum numbers $n_+ = n_- = 0$ and $n_z = 4$, i.e. the quantum number $n_z = 4$ appears for the first time. Looking at higher levels there are maxima of energy differences $\Delta E/\omega_-$ every 5th level above 162 up to level 237. For these levels (162, 167, etc.) the quantum number n_z is 5 and $n_- = 1,2, \ldots$. Between two levels with $n_z = 5$ there are levels with $n_z < 5$ and apparently the energy difference for these is smaller. That is, the difference between harmonic and exact energies is mostly determined by the quantum number n_z. Hence, the anharmonicity of the exact potential is most pronounced in the z-direction. This can also be seen in perturbation theory for higher terms of the expansion of the Coulomb potential where the major contributions to the energy corrections are due to those terms containing high powers of z.

Next let us turn to the wave functions of the hydrogen atom in the well. For the states with either $n_z = 0$ or $n_- = 0$ the expectation values of the y- and z-coordinate are zero. For $n_z = 0$ the wave function is centered near the minimum of the well. With increasing quantum number n_z the center of the wave function is shifted towards negative values of the x coordinate. This is due to the increasing extension of the wave function in the z-coordinate which makes these states "feel" the potential in regions of larger z where the minimum of the well is shifted to negative x-values. We also observe that the extension of the wave functions in x and y is the same for states with the same n_+ and n_- and that the extension in z is the same for states with the same n_z. This shows that excitations in the plane perpendicular to the magnetic field are almost independent from those in the z-direction. Furthermore, in general the extension of the wave functions in the x-y-plane is much smaller than that parallel to the field. Note that the extension of the wave functions in the x-y-plane is also much smaller than it would be in the same potential without the presence of a magnetic field. Apparently, the form of the wave function is determined substantially by the field-dependent kinetic energy.

Looking at the form of the wave functions we restrict ourselves to intersections in the plane perpendicular to the magnetic field. In Fig. 3 we have plotted the square of the absolute value of the wave functions of two states that are highly excited in the plane perpendicular to the magnetic field. The ground state and less highly excited states are not shown since they do not exhibit major differences to the harmonic approximation. Shown are results in the harmonic approximation, this approximation plus corrections from first order perturbation theory for higher terms of the $\frac{1}{|\mathbf{r}|}$-expansion, and the numerical results for the exact wave functions. We see that the harmonic approximation leads to a wave function whose intensity is evenly distributed over an elliptic ring surrounding the minimum of the well. In contrast, due to corrections from anharmonic terms, the intensity of the wavefunctions of the hydrogen atom for states which are highly excited in the plane perpendicular to the magnetic field is largely reduced close to the x-axis both in the perturbation calculation and the exact results. Apparently, this deviation from the wave functions of the harmonic approximation is due to the anharmonicity of the potential. In the perturbation calculation we included only a few terms of the $\frac{1}{|\mathbf{r}|}$- expansion. That

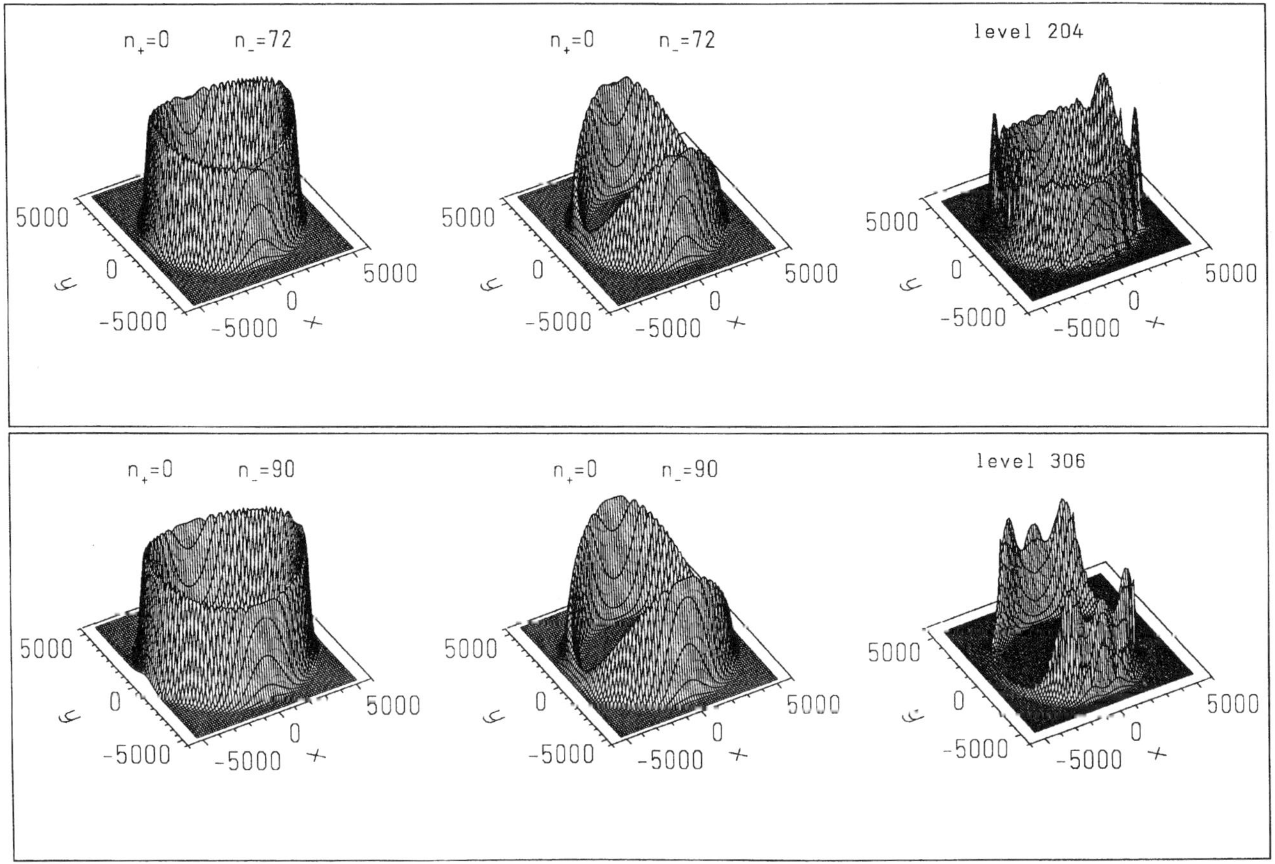

Fig. 3. Square of the absolute value of the wave functions of the hydrogen atom in a magnetic field. Shown are intersections in the plane perpendicular to the magnetic field for the harmonic approximation, for perturbation theory, and for the exact Hamiltonian (from left to right). Parameters: $B = 10^{-5}$, $K = 0.6$ in atomic units

is why the decrease of intensity of the wave functions appears on the level of perturbation theory for even smaller values of the label n_- than in the exact calculation.

Let us conclude with two remarks. First we would like to draw the attention of the reader to the fact that the fundamental aspects discussed in Sect. 2.1 are valid for any neutral two-body system in a magnetic field. The existence of the outer potential well is a universal property of such systems if the pseudomomentum or alternatively the external electric field are sufficiently large. Of particular interest are neutral particle hole systems, i.e. excitons, occurring, for example, in semiconductor bulk systems. Since the Coulomb potential is screened by the ionic background and since the effective masses of the particle and hole are of comparable orders of magnitude, laboratory fields are already strong for the ground state of the exciton. The extension of the bound states in the outer potential well is then of the same order of magnitude as the extension of the well-known states localized in the well due to the Coulomb singularity (see [14]).

Our second remark concerns the experimental observability of the above-mentioned delocalized bound states of the hydrogen atom in crossed electric and magnetic fields. Direct state-to-state transitions for bound states in the outer well should in principle be observable in the radio-frequency regime. A second, probably more promising, approach to experimental verification of the existence of states in the well is the measurement of their dipole moment. There have been experiments that indicate the existence of atoms with very large dipole moments in magnetic fields [15, 16] for energies above the saddle point energy. In [15] results of experiments with Rydberg atoms in magnetic fields ($B \approx 0.1 \cdot 10^{-5}$ a.u.) were published. Their rough estimate for the dipole moment of these atoms is 1.9×10^4 Debye. This value of the dipole moment is roughly of the order of magnitude to be expected.

2.4 Classical Diffusion of the Center of Mass

The subject of investigation of the present section is the classical interaction of the CM and internal motion for neutral two-body systems and in particular for the hydrogen atom in a strong homogeneous magnetic field [17, 18]. For the present we concentrate on the case of a vanishing pseudomomentum $\mathbf{K}$ (the case $\mathbf{K} \neq 0$ will be discussed in Sect. 2.5). The relevant Hamiltonian is then given by Eq. (5) for $\mathbf{K} \equiv 0$:

$$\mathcal{H}_o = \frac{1}{2\mu}\left(\mathbf{p} - e\frac{\mu}{\hat{\mu}}\mathbf{A}(\mathbf{r})\right)^2 + \frac{e^2}{2M}(\mathbf{B} \times \mathbf{r})^2 + V(r). \tag{34}$$

The component of the internal angular momentum parallel to the magnetic field is, for our case of a vanishing electric field as well as pseudomomentum $\mathbf{K} \equiv 0$, a conserved quantity. In spite of the fact that there appear no CM degrees of freedom in the Hamiltonian at Eq. (34) the CM motion is by no means separated from the internal motion. This can be seen by establishing the equations of motion belonging to the Hamiltonian at Eq. (5) and finally putting $\mathbf{K} = 0$. As a result one obtains the equation of motion for the CM

$$\dot{\mathbf{R}} = -\frac{e}{M}(\mathbf{B} \times \mathbf{r}) \tag{35}$$

where $\mathbf{R}$ is the CM coordinate vector. The motional Stark term $-\frac{e}{M}(\mathbf{K} \times \mathbf{B})\mathbf{r}$ in Eq. (5), therefore, intimately couples the CM and internal motion. The CM velocity (see Eq. (35)) is determined by the components of the relative coordinate perpendicular to the magnetic field. It is important to notice that the second quadratic term of the Hamiltonian $\mathscr{H}_o$ represents, according to Eq. (35), the kinetic energy of the CM, i.e. we have $\frac{M}{2}\dot{\mathbf{R}}^2 = \frac{e^2}{2M}(\mathbf{B} \times \mathbf{r})^2$. Since the dynamics of the cyclic CM coordinate is determined by the internal motion, the natural question arises how the transition from regularity to chaos in the internal motion reflects itself in the behaviour of the CM [17, 18]. The phase space of the internal motion is restricted to the energy shell whereas the phase space of the CM motion is, at least in principle, unbounded. Therefore, one may ask whether or not the phase space is filled out by the CM motion depending on the regularity or irregularity of the internal motion.

In the absence of a magnetic field, i.e. for $B = 0$, the pseudomomentum coincides with the total canonical and kinetic momentum. In the field-free space $\mathbf{K} \equiv 0$ means a vanishing CM velocity and the CM stands still. Let us begin our investigation of the CM motion in the presence of a magnetic field with the case of regular internal motion where the Coulomb potential dominates the dynamics. Figure 4 shows a typical trajectory of the CM in the coordinate plane perpendicular to the magnetic field. Since the internal motion is quasiperiodic, the velocity as well as the coordinate of the CM are both also quasiperiodic. Only a bounded part of phase space is, therefore, filled out by the trajectory of the CM. The confinement of the CM trajectories to a circular bounded part of the phase space is a general feature of the deep regular regime. For a complete

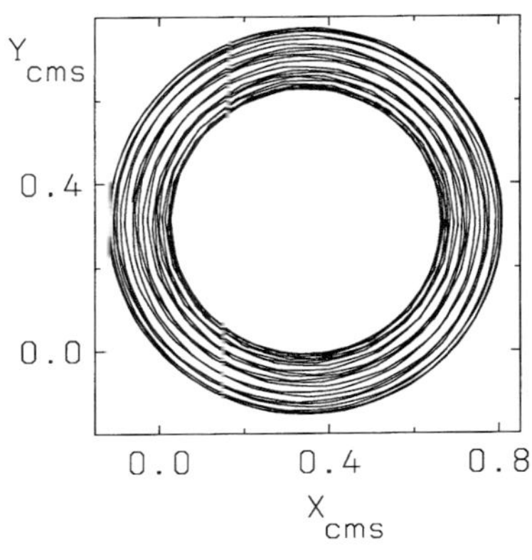

Fig. 4. A typical CM trajectory of the hydrogen atom in the (X_s, Y_s)-hyperplane for regular initial conditions: total energy $E = -10^{-3}$, field strength $B = 10^{-5}$, vanishing pseudomomentum $K = 0$ and angular momentum $L_z = 0$. Starting point is $\mathbf{R} = (0, 0, 0)$. All values in atomic units

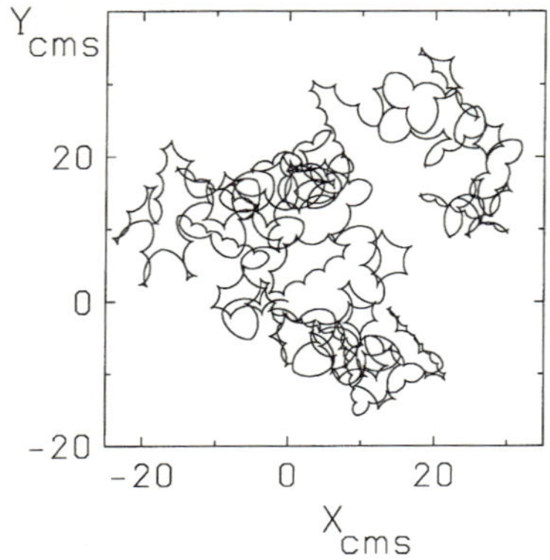

Fig. 5. A typical CM trajectory of the hydrogen atom in the (X_s, Y_s)-hyperplane for chaotic initial conditions: total energy $E = -4.7 \times 10^{-5}$, field strength $B = 10^{-5}$, vanishing pseudomomentum $K = 0$ and angular momentum $L_z = 0$. The time interval of propagation is $T = 1.4 \times 10^8$. Starting point is $\mathbf{R} = (0, 0, 0)$. All values in atomic units

classification of the phase space in the regular regime accessible by low order classical perturbation theory we refer the reader to the literature ([18] contains a classification of the possibilities of the CM motion according to the classification of the internal motion given in [19]).

Next, Fig. 5 shows a typical center of mass trajectory for the case of a full chaotic internal phase space. The eyecatching new feature is that the motion is no more restricted to some bounded volume of phase space. The trajectory of the CM motion of the hydrogen atom in the plane perpendicular to the magnetic field now closely resembles the random motion of a Brownian particle. In fact, the underlying equation of motion at Eq. (35) for the CM motion is a Langevin-type equation without friction. The corresponding stochastic Langevin force is replaced by our intrinsic chaotic force $-e(\mathbf{B} \times \dot{\mathbf{r}})$. A main characteristic of random Brownian motion is the diffusion law, i.e. the linear dependence of the travelled mean-square distance on time. We have plotted in Fig. 6 for our case of a chaotic force for 500 CM trajectories the mean-square distance as a function of time. Within statistical accuracy the plot shows a linear dependence. The mean square distance $\langle \rho_s^2 \rangle$ of the CM after time t, therefore, obeys the diffusion equation

$$\langle \rho_s^2 \rangle = D_s t \tag{36}$$

where D_s is the corresponding diffusion constant. Our intrinsic chaotic force, which is the kicker of the CM motion, therefore possesses the property of randomness in the sense that it provides the well-known diffusion law.

As mentioned above, the energy of the CM motion is contained in a very implicit way in the total Hamiltonian at Eq. (34). In the chaotic regime we have to distinguish carefully between the velocity distribution of the CM and the corresponding diffusion constant resulting from the linear diffusion law. For a typical laboratory magnetic field strength $B = 10^{-5}$ a.u. (≈ 2.35 Tesla) and a binding energy of 5×10^{-5} a.u. which is easily accessible experimentally, we arrive at a typical mean CM velocity of 10^{-5} a.u. ($\approx 20 \frac{m}{s}$). The corresponding diffusion constant is of the order of magnitude of $1.5 \times 10^{-9} \frac{m^2}{s}$. We emphasize

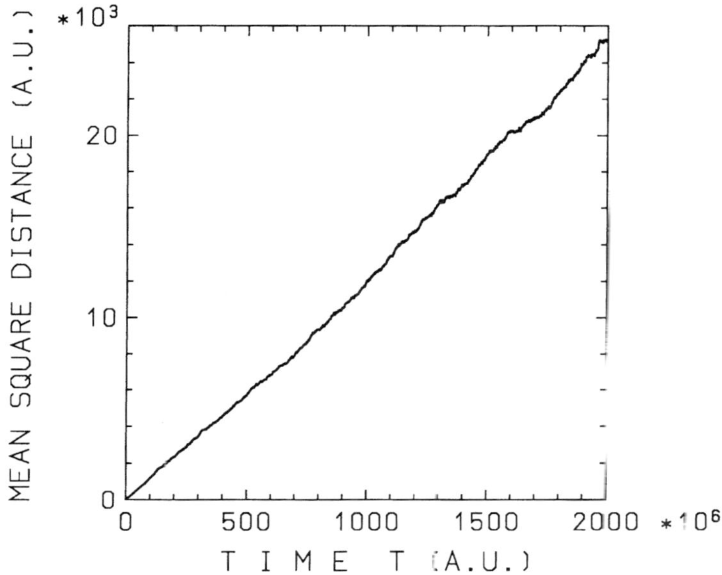

Fig. 6. The mean square distance $\langle \rho_s^2 \rangle$ for an ensemble of 500 trajectories for the hydrogen atom in the fully chaotic regime as a function of time. Parameter values are: $E = -5 \times 10^{-5}$, $B = 10^{-5}$, $K = 0$ and $L_z = 0$. All values in atomic units

that these are results for a vanishing pseudomomentum $\mathbf{K} = 0$ which in the case of the presence of a magnetic field obviously does not mean that the CM velocity is equal to zero. Only the field-free case $\mathbf{K} = 0$ implies that the CM stands still.

Other important examples which exhibit both confinement and diffusion in the classical dynamics of their cyclic collective coordinates are the positronium [20] and the excitonic [14] atom. Because of the comparable masses of the two particles in both cases the mean CM velocity as well as the diffusion constant are orders of magnitude larger than the corresponding values of the hydrogen atom.

For laboratory field strengths the chaotic regime corresponds to highly excited Rydberg states for which the quantum of action is a multiple of the elementary quantum of action $\hbar$. We therefore expect that an approach via classical dynamics is justified and gives some limited insight into the actual physical properties of the atom and, of course, is of interest by its own value. However, it is a priori not evident whether the above observed diffusion of the CM in the chaotic regime survives quantization or whether quantum interference effects will destroy the diffusion, i.e. localize the atom. In addition the effect of quantum localization might depend on the system under consideration (hydrogen atom, positronium, excitons).

2.5 Intermittent Dynamics: a Typical Threshold Phenomenon for Finite Pseudomomentum

In the present section we investigate the classical dynamics of a highly excited neutral two-body system and in particular of the hydrogen atom for a *non-vanishing pseudomomentum* **K** [21]. We will consider energies which are close to the ionization threshold and above the energy of the saddle point. The latter has been established in Sects. 2.1 and 2.3 within the context of our discussion of the potential picture for a neutral two-body system (see also Fig. 1). In Sect. 2.3 we derived an inequality for the absolute value of the pseudomomentum which has to be fulfilled in order to obtain an outer potential well. In the following we assume the existence of the outer potential well, i.e. we consider sufficiently large values of the pseudomomentum or the external electric field.

Let us begin our investigation of the classical dynamics by establishing the equations of motion belonging to the Hamiltonian at Eq. (5):

$$\dot{\mathbf{R}} = \frac{1}{M}\mathbf{K} - \frac{e}{M}(\mathbf{B}\times\mathbf{r}) \tag{37}$$

$$\dot{\mathbf{r}} = \frac{1}{\mu}\mathbf{p} - \frac{e}{2\hat{\mu}}(\mathbf{B}\times\mathbf{r}) \tag{38}$$

$$\dot{\mathbf{p}} = -\frac{e}{M}(\mathbf{B}\times\mathbf{K}) - \frac{\mathrm{e}}{2\hat{\mu}}(\mathbf{B}\times\mathbf{p}) + \frac{e^2}{4\mu}\mathbf{B}\times(\mathbf{B}\times\mathbf{r}) - e^2\frac{\mathbf{r}}{|\mathbf{r}|^3}. \tag{39}$$

In contrast to the case of a vanishing pseudomomentum which has been discussed in the previous section, the pseudomomentum now appears in both the equation for the CM at Eq. (37) as well as the internal equation of motion at Eq. (39). Apart from the purely translational term $(\mathbf{K}/\mathbf{M})t$ the CM motion is again completely determined by the internal coordinate **r**.

The typical new phenomenon for the trajectories of the highly excited hydrogen atom with nonvanishing pseudomomentum $\mathbf{K} \neq 0$ is their intermittent behavior. Intermittency means that the trajectory alternately shows both quasiregular and chaotic phases. Figure 7 shows for a typical trajectory the projection of the internal motion on a plane perpendicular to the magnetic field axis. One immediately realizes that there exist two alternating types of motion. During one phase of motion the electron and the nucleus are in the x, y plane close together and this shows up through the black bubble on the very right of Fig. 7. During this phase of motion the Coulomb and diamagnetic interactions are of comparable order of magnitude and the trajectory is, therefore, chaotic (see [22, 23] for a determination of local Ljapunov exponents). During the other regular looking phase the electron and the nucleus move far apart from each other. The relative motion in the x, y plane then takes place approximately on a circle with a large radius. The Coulomb energy provides here only a small perturbation to the dominating magnetic interaction. The radius of the approx-

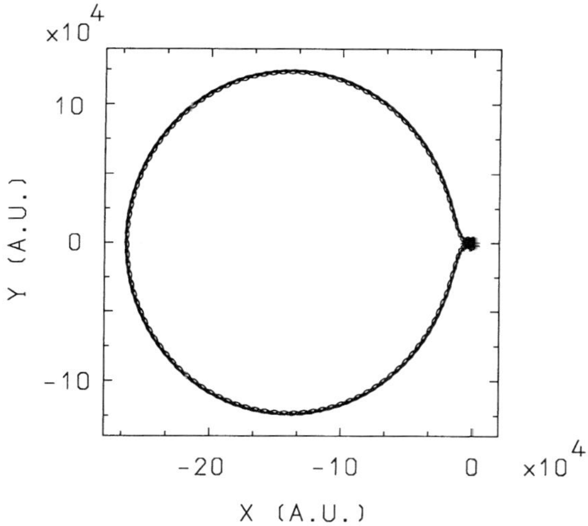

Fig. 7. Typical intermittent trajectory of the internal motion of the hydrogen atom for nonvanishing pseudomomentum (projection onto the plane perpendicular to the magnetic field). Parameter values are: $B = 10^{-5}$, $E = 1.722 \times 10^{-4}$ and $\mathbf{K} = (0, 1, 0)$. All values in atomic units

imate circular large amplitude motion is given by

$$r = -\frac{1}{eB^2}|\mathbf{B} \times \mathbf{K}| \tag{40}$$

i.e. it is completely determined by the field strength and in particular the value of the pseudomomentum. On the other hand, we obtain a completely different interpretation of the pseudomomentum if the electron and the nucleus are very close together. In the latter case the Coulomb dominates the magnetic interaction and hence the pseudomomentum is approximately the linear kinetic momentum of the translational CM motion.

Figure 8 shows the CM motion for the trajectory whose internal motion is given in Fig. 7. It consists of alternating phases of purely translational and circular motions. As already mentioned, the electron and the nucleus are strongly bound, i.e. close together, during the time interval of chaotic internal motion. This is precisely the time period during which the CM performs an almost purely translational motion. The time periods of quasiregular circular internal motion correspond to the periods of circular CM motion. Intermittency, therefore, shows up in the CM motion by alternating phases of more or less straight lined and circular motion.

To complete our picture of the intermittency we have illustrated in Fig. 9 the internal coordinate parallel to the magnetic field as a function of time for the same trajectory whose internal x, y motion and CM motion are shown in Figs. 7 and 8, respectively. During the chaotic phases of the trajectory the z coordinate

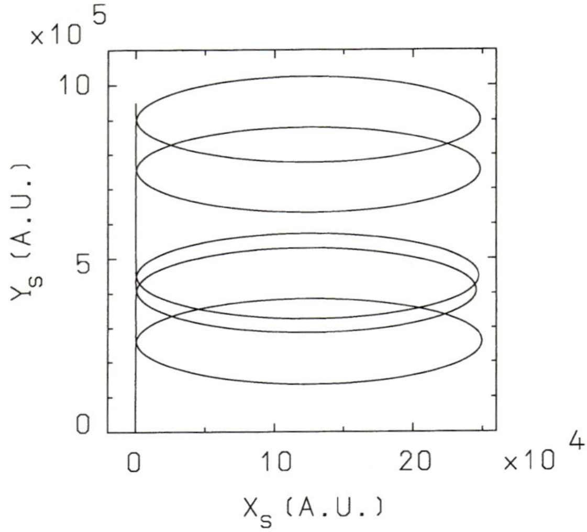

Fig. 8. Center of mass motion of the hydrogen atom belonging to the internal motion shown in Fig. 7. The parameter values are the same as in Fig. 7 and the starting point is $\mathbf{R} = (0, 0, 0)$. All values in atomic units

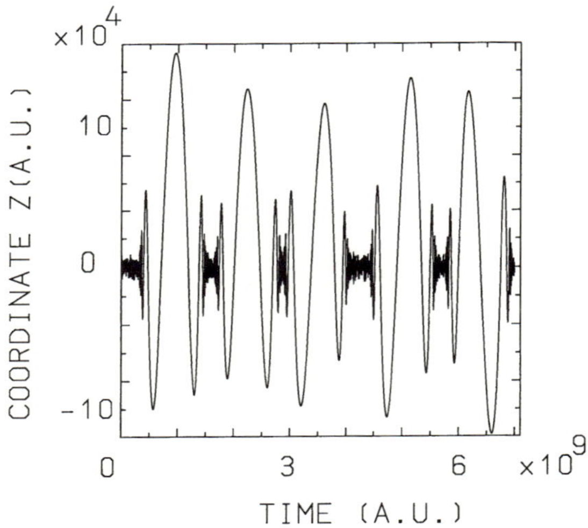

Fig. 9. The motion of the internal z-coordinate as a function of time belonging to the trajectory of the hydrogen atom of Figs. 7 and 8. All values in atomic units

is relatively small which corresponds to the picture that the electron and the nucleus are close together. During the quasiregular phases, i.e. the phases for which the internal x, y motion shows an approximately circular motion with a large radius, the z coordinate shows regular-looking oscillations with large amplitude.

The sudden extension of phase space with increasing energy is a phenomenon which can be easily understood in the context of the potential picture discussed in Sects. 2.1 and 2.3. For energies below the saddle point energy the internal motion is confined either to the singular Coulomb well or to the outer potential well. For energies above the saddle point energy a sudden extension of the available coordinate space takes place and intermittency as a typical phenomenon occurs. The phases of strongly bound motion are in coordinate space located above the Coulomb well whereas the large amplitude motion covers the upper part of the outer potential well.

If we look at the motion of the electron and the nucleus in a plane perpendicular to the magnetic field in the *laboratory coordinate system* we encounter an amazing phenomenon. During the quasiregular phases of motion the electron is localized in a small range of coordinate space whereas the nucleus performs the large amplitude motion on the circle shown in Fig. 7. At first glance this statement seems to contradict the traditional physical picture in which the light electron moves around the heavy nucleus. However, since the Coulomb energy provides only a small perturbation to the magnetic interaction during the quasiregular phase of motion we expect the nucleus and the electron to perform their individual cyclotron motions which are more or less perturbed by the Coulomb interaction. The radius of the cyclotron motion of the nucleus is due to its bigger mass, much larger than that of the electron. The strong localization of the electron and the large amplitude motion of the nucleus is, therefore, a characteristic feature for the case of a strongly dominating magnetic field. In the direction parallel to the magnetic field the large amplitude motion shown in Fig. 9 is performed by the electron.

For a detailed understanding of this phenomenon as well as the dynamical origin of the intermittent behaviour of the trajectories we refer the reader to the literature [21].

3 Charged Two-Body Systems in a Magnetic Field

3.1 Fundamental Properties

The general problem we are concerned with in the present section is a charged two-body system interacting via the Coulomb potential in the presence of an external strong homogeneous magnetic field. The pseudomomentum **K** (see Eq. (2) for $\mathbf{v}_D = 0$) is a conserved quantity for this system and represents a generalization of the CM-momentum in field-free space to the situation in the presence of a magnetic field [11]. In contrast to the case of a neutral system, the two components of the pseudomomentum are not independent for a charged system, i.e. they have a nonvanishing commutator which is proportional to the net charge Q of the system [3, 4, 6]

$$[\mathbf{K}_\alpha, \mathbf{K}_\beta] = -iQ\varepsilon_{\alpha\beta\gamma}\mathbf{B}_\gamma \tag{41}$$

where $\varepsilon_{\alpha\beta\gamma}$ is the Levi-Civita tensor. It is, therefore, for an ion, not possible to eliminate the CM coordinates completely from the Hamiltonian by introducing the pseudomomentum as a canonical conjugated momentum. This is a major difference compared to the case of a neutral system for which such an elimination is possible [3, 4, 6]. Nevertheless, the transformations introduced for a neutral system to perform the so-called pseudoseparation of the CM motion can, in a modified version, also be applied to the case of a charged particle system (see [6, 24] and for a different approach [4]). The resulting transformed Hamiltonian takes on a particularly appealing form and reads for a charged two-body system

$$\mathscr{H} = \mathscr{H}_1 + \mathscr{H}_2 + \mathscr{H}_3 \tag{42}$$

where

$$\mathscr{H}_1 = \frac{1}{2M}\left(\mathbf{P} - \frac{Q}{2}\mathbf{B}\times\mathbf{R}\right)^2 \tag{42a}$$

$$\mathscr{H}_2 = e\frac{\alpha}{M}\left(\mathbf{B}\times\left(\mathbf{P} - \frac{Q}{2}\mathbf{B}\times\mathbf{R}\right)\right)\mathbf{r} \tag{42b}$$

$$\mathscr{H}_3 = \frac{1}{2m}\left(\mathbf{p} - \frac{e}{2}\mathbf{B}\times\mathbf{r} + \frac{Q}{2}\frac{m^2}{M^2}\mathbf{B}\times\mathbf{r}\right)^2 \tag{42c}$$

$$+ \frac{1}{2M_0}\left(\mathbf{p} + \left(\frac{e}{2} - \frac{Q}{2M}\frac{m}{M}(M + M_0)\right)\mathbf{B}\times\mathbf{r}\right)^2 + V$$

where we have already introduced the notation adapted to the case of a one-electron ion, i.e. m, M_0 and M are the electron, nuclear and total mass, respectively. $\alpha = (M_0 + Zm)/M$ and V is the Coulomb potential. For the vector potential $\mathbf{A}$ we have adopted the symmetric gauge. The magnetic field vector $\mathbf{B}$ is again assumed to point along the z-axis. $(\mathbf{R}, \mathbf{P})$ and $(\mathbf{r}, \mathbf{p})$ are the canonical pairs of variables for the CM and relative motion, respectively. The Hamiltonian $\mathscr{H}$ involves five degrees of freedom since the center of mass motion parallel to the magnetic field is a free translational motion, i.e. is separated completely.

The Hamiltonian $\mathscr{H}$ consists of three parts which correspond to different types of motion or interaction. The part $\mathscr{H}_1$ in Eq. (42a), which involves solely the CM degrees of freedom, describes the cyclotron motion of a free pseudoparticle with mass M and charge Q in a homogeneous magnetic field. Within the Hamiltonian $\mathscr{H}_1$ the ion is, therefore, treated as an entity with the net charge and total mass of the ion. This can be looked at as a zeroth order approximation to the real CM motion. We will see in Sects. 3.2 and 3.3 that this zeroth order picture is, in general, not sufficient to describe the CM motion of the ion. In fact the behavior of the CM can deviate strongly from the motion given by the Hamiltonian $\mathcal{H}_1$ [24] and exhibits a variety of different phenomena (see Sect. 2.3 and in particular [25, 26]) depending on the parameter values (energy, field strength, CM velocity). The origin of such rich dynamics lies in particular in the Hamiltonian $\mathcal{H}_2$ in Eq. (42b) which describes the coupling of the CM and electronic degrees of freedom. The Hamiltonian $\mathcal{H}_2$ represents a motional Stark term with a rapidly oscillating electric field which is determined by the dynamics of the system. Because of this "dynamical" electric field the collective and internal motion will, in general, mix up heavily. Finally $\mathcal{H}_3$ in Eq. (42c) contains only the electronic degrees of freedom and describes, to zeroth order, the relative motion of the electron with respect to the nucleus.

An alternative way of writing the Hamiltonian $\mathcal{H}$ is

$$\mathcal{H} = \frac{1}{2M}\left(\mathbf{P} - \frac{Q}{2}\mathbf{B}\times\mathbf{R} - e\alpha\mathbf{B}\times\mathbf{r}\right)^2 + \frac{1}{2\mu}\left(\mathbf{p} - \frac{e}{2M^2}(M_0^2 - Zm^2)\mathbf{B}\times\mathbf{r}\right)^2 + V \tag{43}$$

where $\mu = \frac{mM_0}{M}$ is the reduced mass. This reformulation of the Hamiltonian as a sum of two quadratic terms plus the Coulomb potential gives us some additional insight into the problem. The two Hamiltonian equations of motion

$$\dot{\mathbf{R}} = \frac{1}{M}\left(\mathbf{P} - \frac{Q}{2}\mathbf{B}\times\mathbf{R} - e\alpha\mathbf{B}\times\mathbf{r}\right) \tag{43a}$$

$$\dot{\mathbf{r}} = \frac{1}{\mu}\left(\mathbf{p} - \frac{e}{2M^2}(M_0^2 - Zm^2)\mathbf{B}\times\mathbf{r}\right) \tag{43b}$$

relating to $\mathcal{H}$ immediately allow us to identify the first quadratic term in Eq. (43) as the kinetic energy $E_{cm} = \frac{M}{2}\dot{\mathbf{R}}^2$ of the CM whereas the second quadratic term represents the kinetic energy of the electronic relative motion $\frac{\mu}{2}\dot{\mathbf{r}}^2$. The kinetic energy of the CM, therefore, depends on the electronic degrees of freedom.

For later understanding and interpretation of the dynamics of the ion (see in particular Sect. 3.3) it is also instructive to examine the Newtonian equations of motion. They take on the following appearance:

$$M\ddot{\mathbf{R}} + Q\mathbf{B}\times\dot{\mathbf{R}} + e\alpha\mathbf{B}\times\dot{\mathbf{r}} = 0 \tag{44}$$

$$\mu\ddot{\mathbf{r}} + \frac{e}{M^2}(M_0^2 - Zm^2)\mathbf{B}\times\dot{\mathbf{r}} + e\alpha\mathbf{B}\times\dot{\mathbf{R}} + \frac{\partial V}{\partial\mathbf{r}} = 0. \tag{44a}$$

Again, the mutual dependence of the CM and electronic degrees of freedom becomes obvious. Equation (44) can be integrated once and yields an integration constant which is the pseudomomentum in terms of the CM velocity and the CM and electronic coordinates:

$$\mathbf{K} = M\dot{\mathbf{R}} + Q\mathbf{B}\times\mathbf{R} + e\alpha\mathbf{B}\times\mathbf{r}. \tag{45}$$

Next, let us establish the equations of motion for the CM energy E_{cm} and for the internal energy which is defined as $E_{int} = \frac{\mu}{2}\dot{\mathbf{r}}^2 + V$. Multiplying Eq. (44a) by the CM velocity we obtain the following expression for the time derivative of the CM- and internal energy:

$$\frac{d}{dt}E_{cm} = -\frac{d}{dt}E_{int} = e\alpha(\mathbf{B}\times\dot{\mathbf{R}})\dot{\mathbf{r}}. \tag{46}$$

This equation shows that a permanent flow of energy from the CM to the electronic degrees of freedom and vice versa has to be expected. In Sect. 3.3 the above Eq. (46) will be most helpful for our understanding of the mechanism of the energy transfer between the collective and electronic motion.

3.2 Quantum Mechanical Aspects of the Interaction of the Ionic Center of Mass and Electronic Motions

The present section is devoted to a quantum mechanical investigation of the importance of the Hamiltonian $\mathscr{H}_2$ (see Eq. (42b)) which couples the collective (CM) and internal motion of the ionic system [24]. Because of the "dynamical" electric field which occurs in the Hamiltonian $\mathscr{H}_2$ the collective and internal motion will, as we shall see later, mix up heavily, i.e. it is possible for the ion to change its state of collective or internal motion through the coupling term $\mathscr{H}_2$. Since the ion possesses at least a zero-point Landau energy, the coupling term cannot vanish and is an inherent property of the CM motion of a charged particle system in a magnetic field. This is in contrast to the case of a neutral system where the influence of the CM motion on the internal motion is given by a motional Stark effect with a constant electric field, which can be set equal to zero by choosing the constant pseudomomentum parallel to the magnetic field axis.

In order to investigate the significance and effects of the coupling Hamiltonian $\mathscr{H}_2$ we use in the following a method for formally solving the total Schrödinger equation $\mathscr{H}\mathbf{\Psi} = E\mathbf{\Psi}$. According to our partitioning of the Hamiltonian $\mathscr{H}$ in Eqs. (42) the most natural way is to expand the total wave function $\mathbf{\Psi}$ in a series of products

$$\mathbf{\Psi}(\mathbf{R};\mathbf{r}) = \sum_{p,q} c_{pq}\Phi_p^L(\mathbf{R})\psi_q(\mathbf{r}) \tag{47}$$

where c_{pq} are the coefficients of the product expansion. The functions $\{\Phi_p^L\}$ obey the Schrödinger equation $\mathscr{H}_1\Phi_p^L = E_p^L\Phi_p^L$ for a free particle with charge Q and mass M in a homogeneous magnetic field, i.e. they are the corresponding Landau orbitals. The functions $\{\psi_q\}$ in Eq. (47) are chosen to be eigenfunctions of the electronic Hamiltonian, i.e. $\mathscr{H}_3\psi_q = E_q^I\psi_q$ (q stands collectively for all electronic quantum numbers). If we insert the product expansion at Eq. (47) for the total wave function $\mathbf{\Psi}$ in the total Schrödinger equation and project on a simple product $\Phi_{p'}^L\psi_{q'}$ we arrive at the following set of coupled equations for the coefficients $\{c_{pq}\}$:

$$(\underline{\mathscr{H}_2} + \underline{E^L} + \underline{E^I})\mathbf{c} = E\mathbf{c} \tag{48}$$

where $\mathbf{c}$ is the column vector with components $\{c_{pq}\}$. $\underline{E^L}$ and $\underline{E^I}$ are the diagonal matrices which contain the Landau energies $\{E^L_p\}$ and internal energies $\{E^I_q\}$, respectively. $\underline{\mathscr{H}_2}$ contains the matrix elements of the coupling Hamiltonian $\mathscr{H}_2$. These elements can, via certain commutation relations, be transformed into pure dipole transition matrix elements of the CM as well as internal degrees of freedom (see [24] for details):

$$\underline{\mathscr{H}_2} = -ie\alpha(E^L_{p'} - E^L_p) \cdot \langle \mathbf{\Phi}^L_{p'} | \mathbf{R} | \mathbf{\Phi}^L_p \rangle \cdot [\mathbf{B} \times \langle \psi_{q'} | \mathbf{r} | \psi_q \rangle]\,. \tag{49}$$

Obviously the coupling matrix elements $(\underline{\mathscr{H}_2})_{\{p'q'\}\{pq\}}$ vanish if the two Landau states $\mathbf{\Phi}^L_{p'}$ and $\mathbf{\Phi}^L_p$ possess the same energy $E^L_{p'} = E^L_p$. Transitions, therefore, do occur only for total states which involve with respect to energy different states of the collective motion. The dipole matrix elements between different Landau orbitals of the CM motion (see Eq. (49)) can be calculated analytically in closed form (see [24]). The resulting selection rules are the following. For positive magnetic quantum numbers the transitions involve a change of the magnetic quantum number by one unit and no change of the principal quantum number of the corresponding Landau orbitals. For negative magnetic quantum numbers the transitions take place between Landau orbitals which differ in both the magnetic quantum number as well as the principal quantum number by one unit. Simply stated this means that an appreciable value of the coupling always leads to a strong mixing of different states of collective motion with different energies.

Our original problem of investigating the significance and effects of the couplings between the CM and electronic degrees of freedom is now reduced to the solution of the eigenvalue problem at Eq. (48), i.e. the diagonalization of essentially the coupling matrix $\underline{\mathscr{H}_2}$ in Eq. (49). In the following we will show and discuss different physical situations for which the coupling terms become large at laboratory magnetic field strengths. To estimate the couplings we have to specify our (internal-) electronic wave functions. The simplest way to do this is to choose for the set $\{\psi_q\}$ the eigenfunctions of the hydrogen atom in field-free space and to take into account the electronic diamagnetic interaction via perturbation theory. Such an approach possesses only a limited range of validity with respect to the degree of electronic excitation for which it correctly describes the behavior of the system under consideration. If the ion has developed full (quantum-) chaos, for example, this approach is at least quantitatively inappropriate since the diamagnetic and Coulomb interaction are of comparable orders of magnitude. Nevertheless, as we shall see below, our estimations will allow us to identify regions of strong mixing between the CM and electronic degrees of freedom and will also allow us to draw some qualitative conclusions for the regions where the diamagnetic term in the Hamiltonian $\mathscr{H}_3$ becomes important for the internal motion. For the computational techniques of the evaluation of the dipole matrix elements occurring in Eq. (49) we refer the reader to the literature [24].

In the following we consider the case of a He^+-ion in an external homogeneous magnetic field. The quantities which indicate the relevance of the coupling between the CM and electronic degrees of freedom are not the absolute values κ of the elements of the matrix $\underline{\mathscr{H}_2}$ but rather the quotient of this coupling κ and the energy spacing Δ of the corresponding diagonal matrix elements in $(\underline{E}^L + \underline{E}^I)$. Let us first consider two different electronic states which belong to the same principal quantum number n and couple via the matrix $\underline{\mathscr{H}_2}$. According to the selection rules (see [24]) these two electronic states must differ in both their angular momentum eigenvalues 1 and m by one unit. The corresponding Landau levels of the CM motion differ also by one unit in the magnetic quantum number m_L. Without loss of generality we assume further that $m_L < 0$, i.e. that the Landau levels also differ by one unit in their principal quantum number N.

We begin our discussion with small principal quantum numbers n, for example $n = 2$. With this n-manifold and for strong laboratory magnetic field strengths (≈ 1 Tesla) the Zeeman effect dominates the spin orbit coupling and determines the energy spacing between the electronic states with different eigenvalues m. The diamagnetic interaction is negligible in this range of magnetic field strengths and principal quantum numbers. Our coupling matrix tells us that the typical appearance of the elements κ is

$$\kappa \approx \frac{3e}{MZ} \alpha B [|Q|BN]^{\frac{1}{2}}. \tag{50}$$

The energy spacing Δ is essentially given by the electronic Zeeman energy split $\frac{eB}{2\mu_-}$ with some reduced mass μ_- of the order of magnitude of one. If we assume $N \gg 1$ we arrive at the following estimate of the quotient (κ/Δ):

$$(\kappa/\Delta) \approx \frac{\sqrt{NB}}{M}. \tag{51}$$

For a typical laboratory magnetic field strength $B = 10^{-5}$ a.u. and the mass of the helium ion we need N to be of the order of magnitude $N \approx 6 \times 10^{11}$ in order to make the coupling κ as large as the energy spacing Δ. At first sight this seems to be an astronomical number. However, the energy of the CM motion belonging to this value of N is $E^L = 10^3$ a.u., i.e. some 10 keV. This is a kinetic energy which is easily accessible in the laboratory and still far from the region where relativistic effects start to become significant. Since the number of quantums N in the pure CM Landau motion is very high, the CM motion is expected, apart from small fluctuations, to be well described by its classical path. The influence of the internal motion on the collective motion in this special case is therefore expected to be negligible.

Let us now investigate the behavior of the coupling terms for "intermediate" values of the principal quantum number n. "Intermediate" here means that the Zeeman energy is still dominant over the diamagnetic interaction energies. Since the diamagnetic energies are proportional to B^2n^4 we are, for n up to approximately 15 and within the laboratory achievable magnetic field strengths, well within the above-mentioned regime. The energy spacing for two electronic states of the same n-manifold is then again determined by the Zeeman energy difference. Let us concentrate on the case $n \gg 1$. The order of magnitude of the quantity (κ/Δ) is then determined by the expression

$$(\kappa/\Delta) \approx \frac{\sqrt{NB}}{M} n^2 . \tag{52}$$

As an example we take $n = 10$, $B = 10^{-5}$ a.u. for the He^+-ion. If we again demand that the coupling κ is as large as the energy spacing Δ we arrive at a corresponding value of $N \approx 10^8$, i.e. a kinetic energy of the CM of few ten eV. We conclude therefore that for typical strong laboratory magnetic field strengths, the coupling between the collective and internal motion for states within a higher n manifold are even important for a CM energy of the ion of only a few eV.

Next we consider the coupling of states belonging to different n-manifolds. In order to obtain an appreciable value of the coupling elements we have to go up in the degree of electronic excitation. The energy difference between the levels in the absence of a magnetic field is for large n approximately $\Delta \approx (-Z^2/n^3)$. If our transitions obey the conditions $n \gg l$ and $|n - n'| \ll n$ we obtain [24] the following order of magnitude for our quantity (κ/Δ):

$$(\kappa/\Delta) \approx \frac{\sqrt{NB}}{M} B n^5 . \tag{53}$$

For a magnetic field strength $B = 10^{-4}$ a.u. and $n = 10$ we obtain, from our requirement that κ should be of the order of magnitude of Δ, that $N \approx 10^{10}$, i.e. a CM energy of the ion of a few keV. This means that at these energies the couplings become not only dominant for states within the same n manifold but also important for states belonging to adjacent n manifolds. The exact eigenfunction of the ion can therefore be decomposed into a sum of products of collective and internal wave functions which mixes the Landau orbitals of a certain range of the magnetic and principal quantum numbers and the electronic wave functions over a range of their quantum numbers n, l and m according to the above-mentioned selection rules.

The physical situation discussed so far changes dramatically if we go to higher values of the principal quantum number n. The diamagnetic term of the internal motion becomes more and more important and therefore our description of the electronic wave function in terms of slightly perturbed hydrogen like functions becomes inadequate. The n manifolds are no more well defined, i.e. they mix strongly and the internal wave function continuously changes its shape from initially Coulombic to finally Landau character. These changes are accompanied by a transition from regularity to irregularity, i.e. the onset of quantum chaos [1]. The region where the internal motion is chaotic is also of special interest to our problem of the coupled ionic CM and internal motion (see in particular Sect. 3.3 which contains a classical investigation of the interaction of the ionic CM and electronic motion). Since the energy spacing of the electronic eigenstates becomes much smaller than the corresponding spacing of the levels in the field-free case and since the coupling between the electronic states is probably rather underestimated by the above considerations we expect to obtain a strong mixture of the collective and internal motion, i.e. strong couplings, already for $N \approx 10^4$. As a consequence the mixing of many different states of the CM motion becomes essential for the collective motion and provides no more small fluctuations around a more or less classical CM orbit. We remark that this strong mixing appears automatically because of the finite temperature of the system which corresponds to a certain motional kinetic CM energy ($N \approx 10^4$ is equivalent to $T \approx 5K$).

Let us briefly summarize the main results of the present section. For ions the coupling of the CM and electronic motion in the presence of a magnetic field represents a Stark term with a rapidly oscillating intrinsic dynamical electric field. The Landau orbit of the CM and electronic motion are, therefore, in general intimately coupled. We have investigated different physical situations for atomic ions in a magnetic field for which the interaction between the collective and electronic degrees of freedom becomes strong at laboratory magnetic field strengths. In particular, we have found that the Landau orbit itself can change substantially upon this interaction.

Finally, a comment concerning the relevance of the interaction of the CM and electronic motion for ionic systems in astrophysical magnetic fields is appropriate. For astrophysical field strengths the magnetic energies become, for the ground state of the electronic system, i.e. atom or molecule, comparable to the Coulomb binding energies. Recent investigations ([27] and in particular [28]) have shown for the example of the hydrogen molecular ion that the inherent coupling between the CM and electronic motion is important for a sufficiently strong field even for low-lying electronic as well as CM Landau states of the ion and has a strong impact on the overall structure of the ion.

3.3 *The Classical Self-Stabilization and Self-Ionization Processes*

The aim of the present section is to show and discuss some of the interesting effects and phenomena which arise in the classical dynamics of the interaction of the CM and electronic motion for a charged two-body system [25, 26]. Our starting points are the Newtonian equations of motion for the He^+-ion in a homogeneous magnetic field given in Eqs. (44). In order to solve these coupled nonlinear differential equations for long integration times with a high accuracy one has to regularize the CM and the internal degrees of freedom. For both the analytical regularization transformations as well as the subsequently applied computational techniques we refer the reader to the literature [25, 26, 29].

There are several characteristic regimes of the motion of the He^+-ion in a magnetic field of which we describe here solely the most interesting three. Let us first investigate the regime for which the complete phase space is regular. The internal energy and field strength are chosen such that the Coulomb potential dominates strongly over the magnetic interaction terms. In particular we concentrate on the subset of initial conditions with vanishing CM velocity, i.e. $|\dot{\mathbf{R}}(0)| = 0$. In the complete absence of a magnetic field the ion would simply stay at rest. In the presence of a magnetic field, however, the coupling term $\mathscr{H}_2$ in Eq. (42b) induces an oscillating flow of energy between the CM and internal degrees of freedom which is governed by the energy exchange equation at Eq. (46). The latter equation tells us that extreme values of the CM energy occur if the components of the CM and internal velocities perpendicular to the magnetic field are parallel, i.e. $\dot{\mathbf{r}}_\perp \| \dot{\mathbf{R}}$, whereas a strong flow of energy occurs for the orthogonal configuration, i.e. $\dot{\mathbf{r}}_\perp \perp \dot{\mathbf{R}}$. Typically we now observe four, by orders of magnitude, different time scales for the CM motion. The shortest time scale is that of a single oscillation of the CM energy and motion which corresponds to one slightly perturbed Kepler cycle in the internal motion. The second time scale arises due to the electronic Zeeman term which causes a rotation of the perturbed Kepler ellipses. The third time scale occurs because of the quasiperiodic evolution of the orbital parameters of the ellipses of the internal motion [19] and the action of the coupling Hamiltonian $\mathscr{H}_2$. We find a slow oscillatory modulation of the CM and internal motions (energy) on top of the above-mentioned faster motions. Finally on the fourth and longest time scale the CM performs a circular motion which can be shown to be the motion of a free pseudoparticle with charge Q and mass M in a magnetic field with Larmor frequency $\omega_L = \frac{QB}{M}$. In spite of the fact that the initial CM velocity of the ion was equal to zero we encounter on the longest time scale the effect of self-stabilization of the ion on a Landau orbit. The natural question now arises about the Larmor radius of this orbit. Since we refer to a pseudoparticle picture (which is reasonable if the Coulomb potential dominates the magnetic interaction) the pseudomomentum gives us the coordinates of the center of the Landau orbit of the CM. With the help of the equations of motion we arrive at the following expression for the Larmor radius R_L:

$$R_L = |\frac{e}{Q} \varkappa \mathbf{r}_\perp(0)| . \tag{54}$$

This means that the Larmor radius of the CM motion is completely determined by the initial distance between the electron and the nucleus in the plane perpendicular to the magnetic field. All amplitudes of the oscillations on the above-mentioned shorter time scales are small compared to this Larmor radius. We note that the above-mentioned effect of the classical self-stabilization of the ion on a Landau orbit is a generic phenomenon for regular phase space, i.e. it occurs for any regular initial conditions.

Next, we turn to the regime of motion where the *internal* phase space would be completely chaotic if the nuclear mass were infinite. The initial internal energy and/or the field strength are chosen to fulfill this condition. The initial CM velocity is chosen such that the total energy is close to the threshold $E = 0$. As a characteristic phenomenon of the resulting classical dynamics we observe intermittent behavior [21, 26] of the CM as well as internal motion. After an initial phase of chaotic motion (with a nonvanishing local Ljapunov exponent [22, 23]) and an oscillating flow of energy from and to the CM (internal) motion a sudden strong energy transfer from the CM to the internal degrees of freedom takes place. As a consequence the available phase space for the internal motion is enlarged and a quasiregular phase of large amplitude internal motion follows, during which the magnetic interaction dominates the Coulomb potential. The quasiregular phase ends with a sudden energy transfer from the internal degrees of freedom back to the CM motion and consequently another chaotic phase of motion takes place. This phenomenon occurs repeatedly.

The most interesting dynamics of the ion occurs if we increase the values for the initial CM kinetic energy. The initial internal energy again corresponds to a bound Rydberg state of the ion whereas the total energy is now considerably above zero. After a transient time of bound oscillations in the internal motion (energy) a strong flow of energy from the CM to the internal motion takes place. The *internal* energy is hereby increased above the threshold for ionization, $E_{int} = 0$, and the ion eventually ionizes, i.e. the electron escapes in the direction parallel to the magnetic field. Note that the motion of the electron is confined in the direction perpendicular to the magnetic field.

Figure 10 provides a prototype example for such an ionizing trajectory. Figures 10a and 10b illustrate the time-dependencies of the CM-energy and the z-component of the internal relative coordinate, respectively. After the above-mentioned initial phase of oscillations there occurs at approximately $T = 7 \times 10^6$ a.u. (1.7×10^{-10} s) a sudden loss of CM kinetic energy simultaneously accompanied by an increase in the internal energy which causes the electron to move away from the nucleus in the positive z-direction. The transferred energy, which is in our case of Fig. 10 approximately 6×10^{-3} a.u. (0.2 eV), corresponds only to a small fraction of the total initial CM energy which is for our example about 12.2677 a.u. (333.8 eV). This energy transfer is only possible due to the presence of the coupling term $\mathscr{H}_2$ in the Hamiltonian at Eq. (42) which involves both the internal and CM degrees of freedom. The ionization time for an individual trajectory depends, apart from its intrinsic dynamics, on the field strength and in particular on the CM kinetic energy of the ion.

In order to obtain a statistical measure for the ionization process we have calculated for an ensemble of trajectories the fraction of ionized orbits as a function of time. The initial internal energy was chosen to correspond to a completely chaotic phase space of the He^+-ion if the nuclear mass were infinite. The initial conditions for the internal motion have been selected randomly on the energy shell. In Fig. 11 we have illustrated the fraction of ionized orbits as a function of time up to $T = 10^{10}$ a.u. for a series of different CM energies and for a very strong laboratory field strength of $B = 10^{-4}$ a.u.. For an initial CM energy of $E_{cm} = 0.053$ a.u. which corresponds to an initial CM

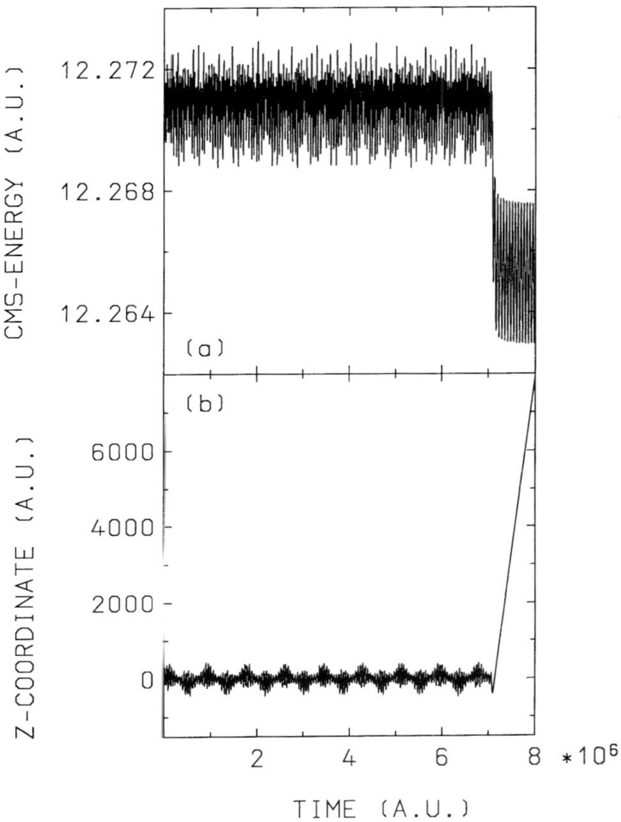

Fig. 10a. Behavior of the CM energy for the He^+-ion as a function of time. **b** Relative coordinate component parallel to the magnetic field as a function of time. The parameter values are: the internal energy $E_{int} = -3 \times 10^{-3}$, field strength $B = 10^{-4}$, kinetic energy of the CM $E_{cm} = 1.581 \times 10^{-2}$ a.u. All values are given in atomic units

velocity of $V_{cm} = 8.4 \times 10^3 \frac{m}{s}$ about 70% of the trajectories are ionized within a time of $T = 10^9$ a.u. $(2.4 \times 10^{-8}$ s) which is the tenth part of the integration time. In contrast to this we have for $E_{cm} = 0.01$ a.u. only about 30% of ionized orbits within the total integration time of $T = 10^{10}$ a.u. $(2.4 \times 10^{-7}$ s). The ionization process depends, therefore, very sensitively on the initial CM kinetic energy of the ion.

All the considered values for the initial internal energy correspond to highly excited Rydberg states of the He^+-ion in a strong magnetic field. According to our quantum mechanical considerations in Sect. 3.2 the perturbation due to the coupling Hamiltonian $\mathscr{H}_2$ becomes, for sufficiently high excited electronic states, larger than the spacing of adjacent levels of the internal Hamiltonian $\mathscr{H}_3$ [24]. As a consequence, strong mixing of the electronic and CM wave functions occurs. This quantum regime of mixing includes the above-mentioned classical

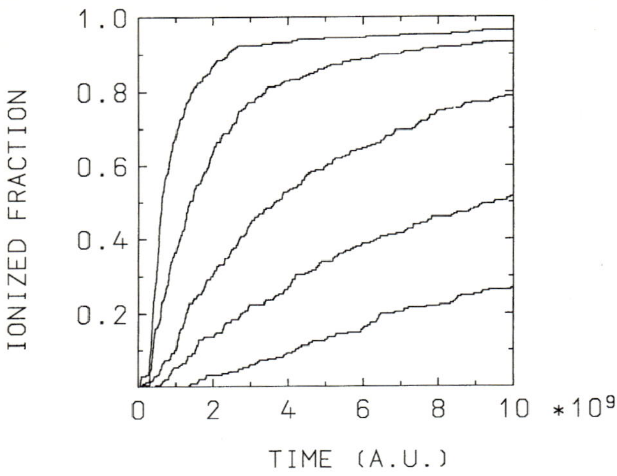

Fig. 11. The ionized fraction for an ensemble of 250 trajectories of the He^+-ion as a function of time. From top to bottom the CM-energies belonging to the ionization curves are $E_{cm} = 5.3 \times 10^{-2}, 2.3 \times 10^{-2}, 1.7 \times 10^{-2}, 1.25 \times 10^{-2}$ and 10^{-2} a.u., respectively. The initial internal energy is always $E_{int} = -3.4 \times 10^{-4}$ a.u. The field strength is $B = 10^{-4}$. All values are given in atomic units

regime for which we observe the process of self-ionization of the ion. Since we are dealing with highly excited states for which the action is much larger than the elementary quantum of action, we expect the self-ionization mechanism of the ion to survive quantization.

4 Summary and Conclusions

We have investigated and discussed a variety of two-body phenomena which arise in the dynamics of neutral and charged two-body systems in strong external fields. The fundamental origin of these effects in the nonseparability of the collective (center of mass) and internal motion of the two interacting particles in the presence of a homogeneous magnetic field. Depending on the net charge of the system the pseudoseparated Hamiltonian takes on a very different appearance. For neutral two-body systems like the hydrogen atom, positronium or an exciton in a semiconductor bulk, a correct treatment of the center of mass yields an outer potential well for the internal motion. This outer well contains bound states which differ from the conventional bound states in the inner singular Coulomb well by their large electric dipole moment and their very high density. The corresponding transitions belong to the radiofrequency regime. Recent experiments on the hydrogen atom in crossed electric and magnetic fields [15, 16] indicate that there exists weakly bound states with a large electric dipole moment which is of the order of magnitude to be expected by the theoretical predictions. These experiments were performed for energies above the saddle point energy of the mentioned outer well. An even greater experimental challenge would be to prepare and investigate states (below the saddle point energy) in the outer potential well.

As a consequence of the collective motion of the neutral system across the homogeneous magnetic field, a motional Stark term with a constant electric field arises. This Stark term inherently couples the center of mass and internal degrees of freedom and hence any change of the internal dynamics leaves its fingerprints on the dynamics of the center of mass. In particular the transition from regularity to chaos in the classical dynamics of the internal motion is accompanied in the center of mass motion by a transition from bounded oscillations to an unbounded diffusional motion. Since these observations are based on classical dynamics, it is a priori not clear whether the observed classical diffusion will survive quantization. From both the theoretical as well as experimental point of view a challenging question is therefore whether quantum interference effects will lead to a suppression of the diffusional motion, i.e. to quantum localization, or not.

For charged-two body systems the interaction of the ionic center of mass and (for an atom) electronic motion is even more intricate. The pseudoseparated Hamiltonian couples the center of mass and internal motion via a motional Stark term which contains a rapidly oscillating electric field of intrinsic dynamical origin. If this coupling becomes relevant, the "dynamical" electric field causes a strong mixing of the Landau orbitals of the center of mass motion and the electronic wave functions. There appears to be a number of different physical situations, i.e. different degrees of electronic excitation and motional states of the center of mass, for which such a mixing occurs. A thorough investigation of the classical dynamics of the He^+-ion reveals a variety of phenomena which occur in the regimes of parameter values where a strong quantum mechanical mixing of the CM and electronic wave functions can be observed. The two major effects are the self-stabilization and the self-ionization processes of the ion. Of particular relevance is hereby the classical self-ionization effect: for large values of the initial center of mass velocity, the energy transfer from the center of mass to the electronic degrees of freedom becomes strong enough in order to allow the atom to ionize.

This could have implications on different physical situations for which the stability of highly excited charged atoms in strong magnetic fields (at finite temperatures) is a relevant question. In particular, the self-ionization phenomenon should be observable in laboratory experiments. In order to detect the energy transfer from the center of mass to the electronic states of the atomic ion under consideration, the following experiment is suggested: a fast beam of atomic ions has to be injected in a homogeneous magnetic field, and subsequently the ions are excited by photons whose appropriately chosen frequency is below the threshold energy for ionization. Nevertheless, ionization, i.e. electron emission in the direction parallel to the magnetic field, should be observable by transfer of energy from the CM to the electronic motion. We note that the typical electromagnetic decay time of the Rydberg states of the ion is of the same order of magnitude as our typical ionization times. They are therefore competing processes. For stronger fields and/or larger CM velocities the self-ionization process dominates.

Acknowledgements. Many valuable discussions with H.D. Meyer and U. Kappes are gratefully acknowledged.

5 References

1. Friedrich H, Wintgen D (1989) Phys Rep 183: 37
2. Delande D, Bommier A, Gay JC (1991) Phys Rev Lett 66: 141
3. Avron JE, Herbst IW, Simon B (1978) Ann Phys (NY) 114: 431
4. Johnson BR, Hirschfelder JO, Yang KH (1983) Rev Mod Phys 55: 109
5. Jackson JD (1983) Klassische Elektrodynamik Aufl. de Gruyter, Berlin
6. Schmelcher P, Cederbaum LS, Kappes U (1994) Kluwer Academic Publishers, Conceptual Trends in Quantum Chemistry, 1–51
7. Dippel O, Schmelcher P, Cederbaum LS (1994) Phys Rev A 49: 4415
8. Schmelcher P, Cederbaum LS (1993) Chem Phys Lett 208: 548
9. Baye D, Clerbaux N, Vincke M (1992) Phys Lett A 166: 135
10. Dzyaloshinskii I (1992) Phys Lett A 165: 69
11. Herold H, Ruder H, Wunner G (1981) J Phys B 14: 751
12. Meyer HD, Kucar J, Cederbaum LS (1988) J Math Phys (NY) 29: 1417
13. Morse PM, Feshbach H, Methods of Theoretical Physics (Mc-Graw Hill, New York, 1953)
14. Schmelcher P (1993) Phys Rev B 48: 14642
15. Fauth M, Walter H, Werner E (1987) Z Phys D 7: 293
16. Raithel G, Fauth M, Walter H (1993) Phys Rev A 47: 419
17. Schmelcher P, Cederbaum LS (1992) Phys Lett A 164: 305
18. Schmelcher P, Cederbaum LS (1992) Zeitschr.f.Phys D 24: 311
19. Delos JB, Knudson SK, Noid DW (1983) Phys Rev A 28: 7 (1983)
20. Schmelcher (1992) J Phys B 25: 2697
21. Schmelcher P, Cederbaum LS (1993) Phys Rev A 47: 2634
22. Fujisaka H (1983) Progr Theor Phys 70: 1264
23. Grassberger P, Procaccia I (1984) Physica (Amsterdam) 13D: 34
24. Schmelcher P, Cederbaum LS (1991) Phys Rev A 43: 287
25. Schmelcher P, Cederbaum LS (1995) Phys Rev Lett 74: 662
26. Schmelcher P (1995) Phys Rev A 52: 130
27. Kappes U, Schmelcher P (1994) Phys Rev A 50: 3775
28. Kappes U, Schmelcher P (1995) Phys Rev A 51: 4542
29. Stiefel EL, Scheifele G (1971) Linear and Regular Celestial Mechanics, Springer

Semiclassical Theory of Atoms and Molecules in Intense External Fields

N. H. March

Oxford University, Oxford, England

To afford a basis for constructing Thomas–Fermi-like approximations to the electronic structure of atoms in intense applied fields, the canonical density matrix (or equivalently the Feynman propagator) for free electrons is first set up. This is done for both intense magnetic and intense electric fields: in each case, exact results are available for arbitrary static field strengths.

Physical properties of atoms and ions in intense magnetic fields are hence obtained in the statistical limit of Thomas–Fermi theory. This discussion is then supplemented by the 'hyperstrong' limit, considered especially by Lieb and co-workers. Chemistry in intense magnetic fields is thereby compared and contrasted with terrestrial chemistry. Some emphasis is then placed on a model of confined atoms in intense electric fields: the statistical Thomas–Fermi approximation again being the central tool employed.

1 Introduction 65

2 One-Electron Hamiltonian and Canonical Density Matrix 65

3 Canonical Density Matrix for Free Electrons in Uniform Magnetic Field of Arbitrary Strength 67

4 Density Matrix for Free Electrons in a Uniform Electric Field 68
4.1 Equation of Motion 68
4.2 Application to Free Electron in Electric Field 69
4.3 Differential Equation for Slater Sum 70

5 Hydrogen Atom Theory 70
5.1 Circular Rydberg States of H Atom in a Magnetic Field 70
5.1.1 Interpolation Formula for H Atom in Magnetic Field Between the Coulomb Result and the Landau High Field Limit 71
5.2 Semiclassical Limit in Uniform Electric Field 72

6 Heavy Atomic Ions in Intense Magnetic Fields 74
6.1 Ground-State Energy Scaling Properties of Heavy Positive Atomic Ions in Extreme High Field Limit 74
6.2 Differential Equation for Density Amplitude $\{\rho(\mathbf{r})\}^{1/2}$ and Concept of Pauli Potential 75
6.3 Delineation of Different Regimes for Heavy (Non-Relativistic) Atoms in Magnetic Fields 76
6.4 Relativistic Semiclassical Theory of Heavy Atoms in an Intense Magnetic Field and the Role of the Virial 77
6.4.1 Relativistic Chemical Potential Equation 79

Private address: Emer Professor, 6, Northcroft Road, Egham, Surrey, TW 20 ODU, England.

7 Model of Confined Atoms in Arbitrary Static Electric Field 80
7.1 Confining Harmonic Force only along Field Direction 81
7.2 Additional Harmonic Force Confinement in x and y Directions 81
7.3 Introduction of Model Potential Energy $V(r)$ Representing Atomic Ion 82
7.3.1 Transcending Thomas–Fermi Approximation 83
7.4 Connection of Harmonic Confining Force Constants and Frequencies with Plasma Density 83
7.5 Simple Illustrative Examples 84

8 Time-Dependent Uniform Electric Field Propagator 87

9 Molecular Ions in Intense Magnetic Fields 87
9.1 H_2^+ Ion: Field along Internuclear Axis 88
9.2 Molecular Anions: Terrestrial and in Intense Magnetic Fields 88

10 Summary and Future Directions 89

Appendix A Effective Potential and Green Function 90

Appendix B Considerations on Current Density 91

References 95

1 Introduction

The discovery of very strong magnetic fields in compact galactic objects, namely white dwarf stars (see, for instance, the review of Garstang [1]) and neutron stars ([2]: see also [3] and references therein) has led to substantial efforts to understand the electronic structure of atoms subjected to such fields.

While a great deal of progress has proved possible for the case of the hydrogen atom by direct solution of the Schrödinger wave equation, some of which will be summarized below, at the time of writing the treatment of many-electron atoms necessitates a simpler approach. This is afforded by the semi-classical Thomas–Fermi theory [4–6], the first explicit form of what today is termed density functional theory [7, 8]. We shall summarize below the work of Hill et al. [9], who solved the Thomas–Fermi (TF) equation for heavy positive ions in the limit of extremely strong magnetic fields. This will lead naturally into the formulation of relativistic Thomas–Fermi (TF) theory [10] and to a discussion of the role of the virial in this approximate theory [11].

A model of confined atoms in an arbitrary static electric field, which can also be solved analytically, will then be discussed in some detail. Contact will be made with results on atomic ions in non-degenerate plasmas, with illustrative examples being presented. A brief treatment follows of the time-dependent uniform electric field Feynman propagator.

Because of interest at the time of writing, molecular ions in intense magnetic fields are treated briefly, with a comparison between terrestrial behaviour and electronic structure in intense galactic magnetic fields. In one of the Appendices, some related considerations on current density in atoms in magnetic fields will be studied.

2 One-Electron Hamiltonian and Canonical Density Matrix

We emphasize from the outset that, throughout the present section, we shall be dealing with the one-electron Hamiltonian

$$H_{\mathrm{r}} = \left(\frac{\left(\mathbf{p} - \frac{e\mathbf{A}}{c}\right)^2}{2m} \right) + V(\mathbf{r}). \tag{1}$$

Here $\mathbf{p}$ is the momentum operator, $\mathbf{A}$ is the vector potential representing the magnetic field $\mathbf{B}$, m the electron mass. The scalar potential energy $V(\mathbf{r})$ can, when desired, include a contribution eFz due to a constant electric field of magnitude F in the z direction.

Below we shall frequently use as a tool to treat strong magnetic and electric fields the canonical density matrix $C(\mathbf{r}, \mathbf{r}_0, \beta)$. This is defined in terms of the eigenfunctions $\psi_i(\mathbf{r})$ and the corresponding eigenvalues ε_i of the Hamiltonian H_r defined in Eq. (1) as

$$C(\mathbf{r}, \mathbf{r}_0, \beta) = \sum_{\text{all i}} \psi_i(\mathbf{r})\, \psi_i^*(\mathbf{r}_0) \exp(-\beta\varepsilon_i) \colon \beta = \frac{1}{k_B T}. \tag{2}$$

This evidently weights the wave function product with the Boltzmann energy factor. However, as shown in general by March and Murray [12], the case of intermediate degeneracy is completely covered also, since one can calculate the density of electrons from the quantity

$$D(\mathbf{r}, \mathbf{r}_0, \beta, \zeta) = \sum_i \frac{\exp(\zeta - \varepsilon_i)}{1 + \exp(\beta(\zeta - \varepsilon_i)} \psi_i(\mathbf{r})\, \psi_i^*(\mathbf{r}_0) \tag{3}$$

which evidently incorporates the Fermi–Dirac function. D is to be obtained from C via the function Q, where

$$C(\mathbf{r}, \mathbf{r}_0, \beta) = \int_0^\infty Q(\mathbf{r}, \mathbf{r}_0, E) \exp(-\beta E)\, dE, \tag{4}$$

D being then given by [12]

$$D(\mathbf{r}, \mathbf{r}_0, \beta, \zeta) = \int_0^\infty Q(\mathbf{r}, \mathbf{r}_0, E) \frac{1}{1 + \exp\{\beta(E - \zeta)\}} dE. \tag{5}$$

The Dirac density matrix corresponding to temperature $T = 0$, i.e. the case of complete degeneracy, with its diagonal element $\mathbf{r}_0 = \mathbf{r}$ the important ground-state electron density $\rho(\mathbf{r})$, is given by the inverse Laplace transform of C/β. This result is correct in the presence of bound states, provided only that a constant greater than the lowest bound-state energy, ε_0 say, is added to the potential energy.

Let us turn immediately to illustrating the use of the canonical density matrix for treating free electrons in a uniform magnetic field. As follows from the definitions in Eqs. (1) and (2), the canonical density matrix satisfies the so-called Bloch equation, as is readily verified, namely

$$H_r C(\mathbf{r}, \mathbf{r}_0, \beta) = -\frac{\partial C(\mathbf{r}, \mathbf{r}_0, \beta)}{\partial \beta}. \tag{6}$$

This is to be solved subject to the boundary condition that as β tends to zero, due to the orthonormality and completeness of the eigenfunctions ψ_i, C tends to the delta function, i.e.

$$C(\mathbf{r}, \mathbf{r}_0, 0) = \delta(\mathbf{r} - \mathbf{r}_0). \tag{7}$$

Equation (6), with the boundary condition at Eq. (7), can be solved analytically for a number of admittedly simple models. Extensive use will be made of these solutions throughout this section.

3 Canonical Density Matrix for Free Electrons in Uniform Magnetic Field of Arbitrary Strength

In a pioneering paper, Sondheimer and Wilson [13] obtained the canonical density matrix C for free electrons in a uniform magnetic field of strength B, taken along, say, the z axis. Their result, denoted below as $C_{\mathrm{OB}}(\mathbf{r}, \mathbf{r}_0, \beta)$, is

$$C_{\mathrm{OB}}(\mathbf{r}, \mathbf{r}_0, \beta) = \frac{1}{(2\pi\beta)^{3/2}} \frac{\beta B}{(\sinh \beta B)} \exp\{-iB(xy_0 - yx_0)\} \exp\left\{-\frac{B}{2}\coth(\beta B) \times [(x-x_0)^2 + (y-y_0)^2] - \frac{(z-z_0)^2}{2\beta}\right\}. \tag{8}$$

The corresponding Hamiltonian H_{OB} has been chosen in a gauge such that

$$H_{\mathrm{OB}} = \tfrac{1}{2}(-i\nabla + \mathbf{B} \times \mathbf{r})^2. \tag{9}$$

While the result in Eq. (8) depends on the gauge chosen for the vector potential in Eq. (9), its diagonal element, $C(\mathbf{r}, \mathbf{r}, \beta)$, the so-called Slater sum, denoted by $S_{\mathrm{OB}}(\mathbf{r}, \beta)$, does not. Indeed, for free electrons, this becomes independent of position $\mathbf{r}$ and is given, in suitable units, by

$$S_{\mathrm{OB}}(\beta) = \frac{1}{(2\pi\beta)^{3/2}} \frac{\mu B \beta}{\sinh(\mu B \beta)} \tag{10}$$

where $\mu = eh/4\pi\, mc$ is the Bohr magneton. As the magnetic field B tends to zero, Eq. (10) reduces simply to $1/(2\pi\beta)^{3/2}$ which is the well known result for the partition function per unit volume of free electrons.

Pfalzner and March [14] have performed numerically the Laplace transform inversion referred to above to obtain the density $\rho(E)$ from the Slater sum in Eq. (10). Below, we shall rather restrict ourselves to the extreme high field limit of Eq. (10), where analytical progress is again possible. Using units in which the Bohr magneton is put equal to unity, the extreme high field limit amounts to the replacement of the sinh function in Eq. (10) by a single exponential term, to yield

$$S_{\mathrm{OB}}(\beta) = \frac{B \exp(-\beta B)}{\pi^{3/2}(2\beta)^{1/2}}. \tag{11}$$

Using again the Laplace transform relation between density and Slater sum, one readily obtains

$$\left.\begin{aligned} \rho(E) &= (2^{1/2}/\pi^2)(E-B)^{1/2} B, \quad E \geqslant B \\ &= 0, \quad E < B. \end{aligned}\right\} \tag{12}$$

This result forms the basis of the semiclassical Thomas–Fermi theory for atoms in intense magnetic fields. Before turning to this topic, however, let us next consider the counterpart of the above example when the magnetic field is replaced by a uniform electric field.

4 Density Matrix for Free Electrons in a Uniform Electric Field

Returning to the basic one-electron Hamiltonian in Eq. (1), the model solved in this present section, following Jannussis [15] and Harris and Cina [16], corresponds to zero magnetic field, i.e. $A = 0$, and to taking the scalar potential energy $V(\mathbf{r})$ as eFz, due solely to the uniform electric field of strength F along the z axis.

4.1 Equation of Motion

To gain some insight into the shape of the solution for the canonical density matrix, let us consider first a one-dimensional problem of electrons moving in a potential $V(x)$. Writing the Bloch equation (Eq. (6)) for the Hamiltonians H_x and H_{x_0} and subtracting them to remove the derivative, one obtains the so-called equation of motion of the density matrix as

$$\left[\frac{d^2}{dx^2} - \frac{d^2}{dx_0^2}\right] C(x, x_0, \beta) = 2[V(x) - V(x_0)]\, C(x, x_0, \beta). \tag{13}$$

Following early work of March and Young [17], one can usefully introduce sum (ξ) and difference (η) variables as

$$\xi = \frac{x + x_0}{2}, \quad \eta = \frac{x - x_0}{2}. \tag{14}$$

Then Eq. (13) takes the compact form

$$\frac{d^2}{d\xi\, d\eta} C(\xi, \eta, \beta) = 4\eta V'(\xi) C(\xi, \eta, \beta) \tag{15}$$

where $V(\xi + \eta) - V(\xi - \eta)$ which enters Eq. (13) has evidently been approximated by $2\eta V'(\xi)$. This is equivalent to the neglect of third and higher derivatives of the one-body potential: V''', V^{iv} etc. Below, for important examples of special interest in this section, it will be seen that these higher derivatives are, in fact, identically zero. Then, of course, Eq. (15) becomes entirely equivalent to the equation of motion (Eq. (13)).

Equation (15) is now amenable to solution by the method of separation of variables [18]. Thus one writes

$$C = J(\xi) K(\eta). \tag{16}$$

Substitution in Eq. (15) then yields the pair of equations

$$\frac{dJ}{d\xi} = s(\beta) V'(\xi) J \tag{17}$$

and

$$\frac{dK}{d\eta} = -\frac{4\eta K}{s(\beta)}. \tag{18}$$

Here, $s(\beta)$ is playing the role of the usual 'separation constant' in this method of solving partial differential equations. The solution of Eq. (18) is immediately found to have the Gaussian form

$$K(\eta, \beta) = A(\beta) \exp\left(\frac{-2\eta^2}{s(\beta)}\right) \tag{19}$$

where $A(\beta)$ and $s(\beta)$ are as yet undetermined functions of β. From the one-dimensional analogue of the free electron result discussed above, for this limiting case corresponding to $V(x) =$ constant, one has $A(\beta) = (2\pi\beta)^{-1/2}$ and $s(\beta) = \beta$.

While for the translationally invariant case, $V'(\xi) = 0$ in Eq. (17), it follows that $J =$ constant is the appropriate solution in this limit, for $V'(\xi)$ non-zero one can also integrate Eq. (17) to read

$$J(\xi, \beta) = J_0(\beta) \exp(-s(\beta) V(\xi)). \tag{20}$$

Thus it is seen that the J and K parts of Eq. (16) are linked by the important separation function $s(\beta)$. Equations (19) and (20) are immediately applicable to one-dimensional motion in a constant electric field, to which problem we now turn specifically.

4.2 Application to Free Electrons in Electric Field

If we return, for a moment to the Sondheimer-Wilson result at Eq. (8) and switch off the magnetic field B, the resulting density matrix $C_{00}(\mathbf{r}, \mathbf{r}_0, \beta)$ must evidently be translationally invariant, and is, in fact

$$C_{00}(\mathbf{r}, \mathbf{r}_0, \beta) = \frac{1}{(2\pi\beta)^{3/2}} \exp\left\{\frac{-|\mathbf{r} - \mathbf{r}_0|^2}{2\beta}\right\}. \tag{21}$$

Intuitively, Eqs. (16) and (20) together then suggest that, in an electric field, the free-electron density matrix at Eq. (21) must be modified by a factor involving $V(z)$, the potential energy due to the field:

$$V(z) = -eFz. \tag{22}$$

In fact, this, as will be seen later, would be the Thomas–Fermi semiclassical treatment of the effect of $V(z)$, namely

$$C_{\mathrm{TF}}(\mathbf{r}, \mathbf{r}_0, \beta) = C_{00} \exp\left[-\beta V\left(\frac{z + z_0}{2}\right)\right]. \tag{23}$$

This form, as it stands, is only an approximate form of solution of the Bloch equation at Eq. (6) valid for small β and/or weak field strength F. Following

Jannussis [15] and Harris and Cina [16], the full solution, denoted by C_{OF}, is given, using $V(z)$ as in Eq. (22), by

$$C_{OF}(\mathbf{r}, \mathbf{r}_0, \beta) = C_{00} \exp\left[\frac{\beta F}{2}(z + z_0) + \frac{\beta^3 F^2}{24}\right]. \tag{24}$$

This result at Eq. (24), given explicitly, for instance, by Harris and Cina [16], when inserted into the appropriate form of the Bloch equation, is readily verified to be an exact solution, as well as satisfying the boundary condition at Eq. (7). The Bloch equation is evidently a powerful tool for treating problems of applied fields of arbitrary strength.

4.3. Differential Equation for Slater Sum

However, for many purposes, the essential physics required is contained in the Slater sum $S(\mathbf{r}, \beta)$, and therefore Lehmann and March [19] and subsequently Amovilli and March [20] have attempted to study this quantity directly. In particular, Lehmann and March obtain an equation for a potential energy $V(\mathbf{r})$ corresponding to a uniform electric field, in D dimensions. This reads

$$\frac{1}{8}\nabla(\nabla^2 S) - V\,\nabla S - \frac{\partial}{\partial\beta}(\nabla S) + \left(\frac{D}{2} - 1\right) S\nabla V = 0. \tag{25}$$

For $D = 3$, and putting $z_0 = z$ in Eq. (24) to obtain the Slater sum S, use of the explicit form of V in Eq. (22) readily allows one to verify that the diagonal form of Eq. (24) is indeed an exact solution of Eq. (25). Later, Amovilli and March [20] made similar progress on central field problems. It remains of interest to treat atoms in intense electric fields by direct use of the Slater sum rather than by use of the off-diagonal canonical density matrix.

We shall return to these free-electron forms of the canonical density matrix in applied **B** and **F** fields of arbitrary strength below. This will allow us to develop semiclassical theory fully within this framework. However, before doing so, we wish to discuss semiclassical theory in a different way for the H atom in intense fields. This will be done, relatively briefly, in the following section.

5 Hydrogen Atom Theory

5.1 Circular Rydberg States of H Atom in a Magnetic Field

As this paper was nearing completion, the work of Germann et al. [21] appeared. In this work, dimensional perturbation theory has been employed to study circular Rydberg states of the H atom in a magnetic field. These authors note first that dimensional scaling methods that employ the dimensionality D of space as a variable have proved effective in treating non-separable problems which lack a natural expansion parameter [22]. In particular, the large dimensionality limit, $D \to \infty$, is often amenable to exact analysis that can then be utilized to derive $D = 3$ results of useful accuracy from an expansion in powers of $1/D$ [23]. Related expansions have been applied to atoms in external fields (see for example [24]).

It is very relevant at this point to note the important work of Bender et al. [25] on the H atom in a uniform magnetic field. These authors derived a semiclassical expansion for the ground-state energy in powers of $(2|m| + 2)^{-1}$, where m is the azimuthal quantum number.

Since, it turns out, the D and $|m|$ dependences enter the Schrödinger equation only via the factor

$$K = D + 2|m| - 1, \tag{26}$$

dimensional perturbation theory is equivalent to angular momentum perturbation theory about $|m|$ tending to infinity. At dimensionality $D = 3$, this is then identical to the expansion of Bender et al. [25].

The approach of Germann et al. [21], in contrast to other procedures, is applicable to the entire range of magnetic field strengths. Energies and expectation values for circular Rydberg states are presented in this work as functions of the field strength.

5.1.1 Interpolation Formula for H Atom in Magnetic Field between the Coulomb Result and the Landau High Field Limit

Bender et al. [25] show that, by semiclassical expansion, the leading approximation to the ground-state energy E for the H atom in a magnetic field may be written

$$E = \frac{B}{\eta^2}\left[\frac{1}{8}(3\eta^4 - 1)k\right] \tag{27}$$

where $(k/2) = |m| + 1$, m being the azimuthal quantum number. Here η is such that for large magnetic fields it approaches unity as

$$\eta = 1 - \frac{Z}{B^{1/2}k^{3/2}} + \cdots. \tag{28}$$

This same quantity vanishes for low field strengths according to

$$\eta = \frac{B^{1/2}k^{3/2}}{4Z}. \tag{29}$$

Including Gaussian fluctuations around the classical minimum, Bender et al. [25] obtain

$$E = \frac{B}{\eta^2}\left\{\frac{1}{8}(3\eta^4 - 1)k + \frac{1}{4}[(1+3\eta^4)^{1/2} + (1-\eta^4)^{1/2} - 2]\right\} \tag{30}$$

The Coulomb limit $B = 0$ of Eq. (30) is obtained using Eq. (29) as

$$E = -\frac{2Z^2}{k^2} = -\frac{Z^2}{2(|m|+1)^2}. \tag{31}$$

This result coincides with the exact ground-state energy in each azimuthal eigenspace, as is explicitly verified by writing the exact Coulomb level spectrum in terms of parabolic quantum numbers as

$$E_{B=0} = -\frac{Z^2}{2(|m| + n_1 + n_2 + 1)^2} \tag{32}$$

and setting the quantum numbers n_1 and n_2 equal to zero (compare [26]).

The extreme high field limit B tends to infinity is obtained from Eq. (30) by making use of Eq. (28). The result is

$$E = \tfrac{1}{4}\,Bk = \tfrac{1}{2}\,B(|m|+1) \tag{33}$$

which coincides with the exact Landau spectrum for the ground-state energy [25].

The above demonstrates therefore that the Gaussian approximation at Eq. (30) correctly interpolates between the Coulomb and Landau limits. This, needless to say, does not mean that Eq. (30) is exact throughout the entire span of magnetic field intensities. However, it can be taken as encouraging for the utility of the semiclassical expansion, after higher-order corrections have been incorporated.

5.2 Semiclassical Limit in Uniform Electric Field

We referred above to the important semiclassical study of Bender et al. [25] for a H atom in a magnetic field. These writers also give a brief description of the application of their semiclassical expansion to the H atom in a uniform electric field. Suppressing terms that are unimportant for the semiclassical limit, they write the Schrödinger equation in the form

$$\left[-\frac{1}{2}\left(\frac{\partial^2}{\partial\rho^2} + \frac{\partial^2}{\partial z^2}\right) + \frac{k^2}{8\rho^2} - \frac{1}{(\rho^2+z^2)^{1/2}} + Fz\right]\phi = E\phi \tag{34}$$

where F denotes the strength of the electric field directed along the z axis.

Introducing rescaled coordinates u and v through

$$\rho = k^2 u; \quad z = k^2 v, \tag{35}$$

the above Schrödinger equation becomes

$$\left[-\frac{1}{2k^2}\left(\frac{\partial^2}{\partial u^2}+\frac{\partial^2}{\partial v^2}\right)+V(u,v)\right]\phi=(Ek^2)\phi, \tag{36}$$

where

$$V(u,v)=\frac{1}{8u^2}-\frac{1}{(u^2+v^2)^{1/2}}+\bar{F}v \tag{37}$$

with

$$\bar{F}=Fk^4. \tag{38}$$

In the course of the semiclassical development of Bender et al. [25], the effective electric field $\bar{F}$ is treated as a k-independent constant. Then its true value given by Eq. (38) is restored when the calculation has otherwise been completed.

It turns out, in fact, that the large k limit sought leads to an approximation in which the kinetic energy is suppressed and thus the leading contribution to the ground-state energy is determined by the minimum of the effective potential $V(u,v)$. The stationary points of $V=V(u,v)$ are evidently roots of the algebraic equations $\partial V/\partial u=0=\partial V/\partial v$. Explicitly, one then finds that [25]

$$v=-4\bar{F}u^4\colon\ 4u=(1+16\bar{F}u^6)^{3/2}. \tag{39}$$

For a sufficiently weak electric field, the second of these equations has two distinct positive roots $u=u_1$ and $u=u_2$, ordered such that $u_1<u_2$. v is accordingly determined by the first equation. Only the small root u_1 corresponds to a local minimum of $V(u,v)$ so that the semiclassical expansion is performed around $u=u_1$ and $v=v_1=-4\bar{F}u_1^4$. The values u_1 and u_2 approach one another with increasing strength of the electric field and eventually become equal at some $\bar{F}=\bar{F}_0$. *It is then found* [25] that the (real) local minimum disappears for $\bar{F}>\bar{F}_0$. The above treatment of Bender et al. has its counterpart in more conventional terms in the article by Bethe and Salpeter [27].

The value F_0 can be readily calculated and is equal to $2^{16}/3^9$. The field strength corresponding to this is given in V/cm by [25]

$$F_0=\frac{1.07\times 10^9}{(k/2)^4} \tag{40}$$

where $(k/2)=|m|+1$. As could have been anticipated intuitively, the electric field required for ionization of the atom is smaller for large $|m|$, i.e. for high angular momentum states.

Bender et al. [25] stress that the above classical calculation of F_0 neglects both quantum fluctuations and barrier penetration. However, their study points to the possibility of a semiclassical expansion for $F<F_0$.

6 Heavy Atomic Ions in Intense Magnetic Fields

Returning briefly to free electrons in zero field, the Slater sum $S_{00}(\beta) = (2\pi\beta)^{-3/2}$ is readily Laplace inverted to obtain the density $\rho(E) \propto E^{3/2}$ and the density of states $N(E) = d\rho/dE \propto E^{1/2}$. For the case of intense magnetic fields applied to free electrons, $\rho(E)$ is given analytically in Eq. (12). To obtain the Thomas–Fermi theory in this intense field regime, the (Fermi) energy E, put equal to the chemical potential μ, is to be replaced by $\mu - V(\mathbf{r})$, with $V(\mathbf{r})$ the (assumed) slowly varying potential energy. Then we have the extreme high field Thomas–Fermi results for the (weakly) inhomogeneous electron density $\rho(\mathbf{r})$ as

$$\rho(\mathbf{r}) = (2^{1/2}/\pi^2)(\mu - V(\mathbf{r}) - B)^{1/2} B. \tag{41}$$

We can usefully solve this for the chemical potential μ as

$$\mu = \frac{2\pi^4\rho^2}{B^2} + V(\mathbf{r}) \tag{42}$$

where in Eq. (42) it has proved convenient to lump the field B into the redefined chemical potential. Below we shall obtain the non-relativistic kinetic energy density $t(\mathbf{r})$ for electrons in intense magnetic fields, and the first term on the right of Eq. (42) is, in fact, in the local density approximation employed in Thomas–Fermi theory; just $\partial t/\partial\rho$.

6.1 *Ground-State Energy Scaling Properties of Heavy Positive Atomic Ions in Extreme High Field Limit*

In the extreme high field limit, the scaling properties of the ground-state energy $E(Z, N, B)$ can be exposed for positive atomic ions with nuclear charge Ze and with $N \leqslant Z$ electrons from the Thomas–Fermi (TF) theory set out above [28].

To gain orientation, let us first neglect the self-consistent field (see below for its inclusion). The TF solution for the diagonal element of the canonical density matrix or the Slater sum $S(\mathbf{r}, \beta)$ can be written (compare Eq. (20)) as $S_{OB}(\beta)\exp(-\beta V(\mathbf{r}))$ with $V(\mathbf{r}) = -Ze^2/\mathbf{r}$. This is readily shown to yield, via the ground-state electron density $\rho(\mathbf{r}, B) = L^{-1}\{S(\mathbf{r}, \beta, B)/\beta\}$ where L^{-1} denotes the inverse Laplace transform discussed earlier:

$$E_{TF}^{Coul} = \text{constant } Z^{9/5}B^{2/5}(N/Z)^{3/5}. \tag{43}$$

Introducing the self-consistent field, March and Tomishima [28] obtained the generalization of the Coulomb result (6.3) as

$$E_{TF}(Z, N, B) = Z^{9/5}B^{2/5}f(N/Z) \tag{44}$$

where, for small N/Z, $f \propto (N/Z)^{3/5}$ leads back to Eq. (43). These scaling relations, valid as stressed for heavy positive atomic ions, can also be related to the generalized $1/Z$ expansion in an intense magnetic field, extending the work of March and White [29] in zero field.

Numerical self-consistent fields have been established (see, for example [9, 30, 31]) and we shall record below a few of the consequences of the calculations, in particular for the chemical potential μ, when we have considered the relativistic generalization of the TF theory (Sect. 6.4 below).

6.2 Differential Equation for Density Amplitude $\{\rho(\mathbf{r})\}^{1/2}$ *and Concept of Pauli Potential*

Early work of March and Murray [32] (see also [33]) used the so-called von Weizsäcker equation, which is the TF method, corrected by the lowest order density gradient term in the kinetic energy, for the electron density $\rho(\mathbf{r})$. They demonstrated that this density functional theory could be recast into the form of a Schrödinger equation for the ground-state density amplitude $\{\rho(\mathbf{r})\}^{1/2}$, but with an important modification to the one-body potential energy involved in that equation. Some decades later, various research groups converged independently (for references see [33]; see also [6]) on the Schrödinger equation

$$\nabla^2\{\rho(\mathbf{r})\}^{1/2} + \frac{2m}{\hbar^2}[\varepsilon - V(\mathbf{r}) - V_{\text{Pauli}}(\mathbf{r})]\{\rho(\mathbf{r})\}^{1/2} = 0. \tag{45}$$

Here $V(\mathbf{r})$ is the so-called Slater–Kohn–Sham (SKS) potential involved in finding one-electron wave functions $\psi_i(\mathbf{r})$ yielding for an N-electron atom or molecule the ground-state density $\rho(\mathbf{r})$ as

$$\rho(\mathbf{r}) = \sum_{i=1}^{N} \psi_i(\mathbf{r})\psi_i^*(\mathbf{r}). \tag{46}$$

The SKS potential $V(\mathbf{r})$ is formally

$$V(\mathbf{r}) = V_{\text{Hartree}}(\mathbf{r}) + \frac{\delta E_{\text{xc}}[\rho]}{\delta\rho(\mathbf{r})} \tag{47}$$

but, of course, exact knowledge of the exchange-correlation functional $E_{\text{xc}}[\rho]$ involves complete solution of the many-electron problem. While this goal still looks remote at the time of writing, Holas and March [34] have shown that the exchange-correlation potential $V_{\text{xc}}(\mathbf{r})$ adding to the self-consistent Hartree term in Eq. (47) can, in fact, be written quite explicitly in terms of the first- and second-order density matrices introduced by Löwdin [35].

Though the Pauli potential $V_{\text{Pauli}}(\mathbf{r}) \equiv V_{\text{p}}(\mathbf{r})$ entering the Schrödinger equation at Eq. (45) is also only known approximately (for example, the von Weizsäcker study of March and Murray [32] yielded $V_{\text{p}}^{\text{Weiz}}(\mathbf{r}) = (5/3)c_k\{\rho(\mathbf{r})\}^{2/3}$ where the kinetic constant $c_k = (3h^2/10m)(3/8\pi)^{2/3}$), the work of Lieb et al. [36] (see also [37]), who did not utilize the Pauli potential in their original paper, proves to be a splendid example of this approach, where important analytical progress proves possible.

6.3 *Delineation of Different Regimes for Heavy (Non-Relativistic) Atoms in Magnetic Fields*

Before returning to the Pauli potential, it is important to classify five regions, following the work of Lieb et al. [36] on the calculation of the ground-state energy of an atom having nuclear charge Ze in a magnetic field B in the asymptotic regime of very large Z, already considered above in the TF approximation. Lieb et al. [36] usefully classify the following regimes, in this limit of very large Z, which, in suitable units, correspond to (i) $B \ll Z^{4/3}$, (ii) $B \approx Z^{4/3}$, (iii) $Z^{4/3} \ll B \ll Z^3$, (iv) $B \approx Z^3$ and (v) $B \gg Z^3$.

Regions (i), (ii) and (iii) correspond to the TF semiclassical form already discussed above. Region (iv), as Lieb et al. [36] point out, can be treated by a simple density matrix functional theory. It is, however, region (v) which can be discussed very naturally in terms of the Pauli potential introduced above.

This regime is described by Lieb et al. [36] as one in which atoms become 'needles'. If the magnetic field is chosen as the z axis, then these workers characterize the atom by a one-dimensional density $\bar{\rho}(z)$. Apart from some scaling (see [36]), $\bar{\rho}$ is the cross-sectional integral of the three-dimensional density $\rho(\mathbf{r})$. In the above situation, where the atom degenerates into a needle, the Coulomb potential can be replaced by a delta function. Then, writing ψ for the density amplitude $\rho^{-1/2}$, Lieb et al. [36] derive the Schrödinger equation (again in suitable units):

$$\psi''(z) + [\mu + \delta(z) + \psi^2(z)]\psi(z) = 0, \tag{48}$$

which is of precisely the form of the 'Bosonized' Schrödinger equation at Eq. (45). Introducing $\lambda = N/Z$, with N the total number of electrons, Lieb et al. [36] solve for $\psi(z)$ to find

$$\psi(z) = \frac{2^{1/2}(2-\lambda)}{4\sinh[\frac{1}{4}(2-\lambda)|z| + c]}, \lambda < 2 \tag{49}$$

with tanhc $= (2-\lambda)/2$. The ground-state energy is given by

$$E = Z^3[\ln\eta]^2\, E^{\mathrm{HS}}(N/Z)\colon \eta = B/Z^3, \tag{50}$$

where the superscript HS stands for the hyperstrong field regime under discussion where $B \gg Z^3$ in the limit of very large Z. The explicit form of $E^{\mathrm{HS}}(\lambda)$ is derived by Lieb et al. [36] as

$$E^{\mathrm{HS}}(\lambda) = -\tfrac{1}{4}\lambda + \tfrac{1}{8}\lambda^2 - \tfrac{1}{48}\lambda^3. \tag{51}$$

Since $\lambda < 2$ to obtain the above solution, it is remarkable that the maximum number of electrons is now only just less than $2Z$. The neutral TF atom in zero magnetic field cannot bind even one extra electron. Thus the situation is totally changed from that to which we are accustomed in terrestrial chemistry. Not only can N be near to $2Z$, but the binding energy of the last Z electrons is of the same order of magnitude as that for the first Z electrons. Furthermore, the binding energy of two identical atoms is greater than the energy of the two individual atoms. This is in diametric contrast to the TF homonuclear diatomic molecule in zero magnetic field, which the theorem of Teller [38] assures us is not bound! Thus terrestrial chemistry must be completely changed as one approaches the surface of a neutron star (see also Freeman and March [39] and Sect. 9 below).

6.4 Relativistic Semiclassical Theory of Heavy Atoms in an Intense Magnetic Field and the Role of the Virial

The relativistic TF theory for a heavy atom in intense magnetic fields has been investigated numerically by Hill et al. [40], and their main findings are discussed elsewhere in this paper. However, the main purpose of this section is to demonstrate, for any completely local relativistic density functional theory (DFT), that the total kinetic energy can be related to the virial plus contributions from the chemical potential of the electron cloud and classical potential energy terms.

If the relativistic kinetic energy functional $T[\rho]$ is completely local, then one can rewrite the general chemical potential equation of DFT as

$$\mu = \frac{dt}{d\rho} + V(\mathbf{r}) \tag{52}$$

where t is the local kinetic energy density, defined such that

$$T = \int t \, d\mathbf{r}. \tag{53}$$

Now, for simplicity, consider a heavy atom with spherical electron density $\rho(r)$. One can then rewrite Eq. (53) as

$$T = \int t \, 4\pi r^2 \, d\mathbf{r} = -\int \frac{dt}{\mathrm{d}r} \frac{4\pi}{3} r^3 \, dr. \tag{54}$$

But in the completely local theory under consideration, the derivative dt/dr appearing in Eq. (54) can be replaced by $(dt/d\rho) d\rho/dr$. Using the Euler-Lagrange equation of DFT for $dt/d\rho$, one can readily express T in Eq. (54) as

$$T = -\int \frac{4\pi}{3} (\mu - V(r)) \, r^3 \frac{d\rho}{dr} \, dr. \tag{55}$$

Integrating once more by parts, one finds

$$T = N\mu - \int \rho \, V \, d\mathbf{r} - \frac{1}{3} \int \rho \mathrm{r} \frac{dV}{dr} \, dr. \tag{56}$$

This then relates T in the ground state to (i) the chemical potential μ, (ii) the potential-energy contribution $\int \rho \, V \, d\mathbf{r}$ and (iii) the virial $\mathbf{r} . \mathbf{F}$, where the force

$\mathbf{F} = -\operatorname{grad} V$. Thus one can express Eq. (56) as

$$T = N\mu - \int \rho V \, d\mathbf{r} + \frac{1}{3}\langle \mathbf{r} \cdot \mathbf{F} \rangle, \tag{57}$$

where the average $\langle \mathbf{r} \cdot \mathbf{F} \rangle$ simply weights the virial $\mathbf{r} \cdot \mathbf{F}$ with the density $\rho(\mathbf{r})$. It is to be emphasized that, given a completely local theory based on the Euler equation at Eq. (52), the result at Eq. (51) follows quite generally without specifying $t(\rho)$.

In the above context, let us consider the relativistic TF theory of Hill et al. [40] for heavy atoms in an intense magnetic field. The Euler equation now takes the explicit form (see also Eq. (66) below)

$$\mu = \left[\frac{c^2 h^4 \rho^2}{4e^2 B^2} + m_0^2 c^4\right]^{1/2} - m_0 c^2 + V(\mathbf{r}) \tag{58}$$

where m_0 is the rest mass of the electron and c the velocity of light. Hence, for the intense magnetic fields B for which Eq. (58) is valid, comparison with Eq. (52) yields immediately

$$\frac{dt}{d\rho} = \left[\frac{c^2 h^4 \rho^2}{4e^2 B^2} + m_0^2 c^4\right]^{1/2} - m_0 c^2. \tag{59}$$

This is now the point at which to return to Eq. (52). Multiplying by $\rho(\mathbf{r})$ and integrating over the whole of space then yields

$$N\mu = \int \rho \frac{dt}{d\rho} \, d\mathbf{r} + \int \rho V d\mathbf{r}. \tag{60}$$

Using Eq. (60) in Eq. (57) one then finds

$$T = \int \rho \frac{dt}{d\rho} \, d\mathbf{r} + \frac{1}{3}\langle \mathbf{r} \cdot \mathbf{F} \rangle. \tag{61}$$

One can remove the force from Eq. (61) by differentiating the Euler equation at Eq. (52) with respect to r to obtain

$$\frac{dV}{dr} = -\frac{d}{dr}\left(\frac{dt}{d\rho}\right). \tag{62}$$

Then, for the case of spherical symmetry, one can rewrite the average of the virial as

$$\langle \mathbf{r} \cdot \mathbf{F} \rangle = \int \rho r \frac{d}{dr}\left(\frac{dt}{d\rho}\right) d\mathbf{r}. \tag{63}$$

The conclusion therefore is that, within the completely local framework above, one can calculate the kinetic contribution T to the ground-state energy

E directly from the density ρ as

$$T = \int \rho \frac{dt}{d\rho} d\mathbf{r} + \frac{1}{3} \int \rho r \frac{d}{dr}\left(\frac{dt}{d\rho}\right) d\mathbf{r}. \tag{64}$$

Evidently, for the relativistic TF theory of heavy atoms in an intense magnetic field B, one can insert Eq. (59) into Eq. (64) to obtain T explicitly in terms of ρ and its first derivative $d\rho/d\mathbf{r}$. As a limiting case of this result, the non-relativistic expression for the kinetic energy density is readily confirmed to be

$$t_{\text{non-rel}}(\rho) = \frac{h^4 \rho^3}{24 e^2 B^2 m_0}, \tag{65}$$

$3t_{nr}$ resulting from the first term on the right of Eq. (64). The second term must evidently cancel $2t_{nr}$ of this total, and this can be readily verified again after an integration by parts. Equation (65) is, of course, invoked elsewhere in this paper in the discussion of the non-relativistic(nr) semiclassical TF theory of heavy atoms in intense magnetic fields.

6.4.1. Relativistic Chemical Potential Equation

As noted above, the chemical potential equation can be generalized from the non-relativistic form in Eq. (42) to include non-zero fine structure constant α. The result in Eq. (58) can be rewritten as [6, 40]

$$(\mu_\alpha - V)^2 + 2m_0c^2(\mu_\alpha - V) = \frac{c^2 h^4}{4(eB)^2} \rho^2(\mathbf{r}) \tag{66}$$

Table 1. Chemical potential μ of positive atomic ions from TF theory in intense magnetic fields, without (subscript zero) and with (subscript α) the fine structure constant. Taken from Hill et al. (1983, 1985).

Example below is for atomic number $Z = 100$, and for varying degrees of ionization $q = 1 - (N/Z)$ where $N < Z$ is the number of electrons in positive ions of atomic number Z. All results recorded are in an intense magnetic field of strength $B = 10^{14}$G[a]

q	$\mu_0(Z, N)$	$\mu_\alpha(Z, N)$
0.1	−240.7	−241.5
0.2	−535.1	−537.4
0.3	−881.4	−885.9
0.4	−1290	−1297
0.5	−1779	−1792

[a] It will be a straightforward matter to obtain specific results for, say, the Crab pulsar's field, estimated at 6×10^{12}G (see, for instance, [1]) should precise data for given fields be needed subsequently, using TF theory in the limit of large **B**, from which the Table was constructed.

where μ_α denotes the non-relativistic chemical potential. Hill et al. have solved this equation self-consistently by numerical methods. As one example of their results we record in Table 1 a comparison between chemical potentials without (μ_0) and with (μ_α) relativistic corrections. The results shown are for specific magnetic field and Z but for positive atomic ions in different states of ionization. The corrections are not large per centage-wise, from inclusion of relativity. This means that inclusion of the quadratic term in $(\mu_\alpha - V)$ on the left of Eq. (66) does not alter the chemical potential in a major way. Dropping this term, which is correct in the limit c tends to infinity or the fine structure constant α tends to zero, with the rest mass m_0 simply m, Eq. (66) in suitable units reverts to the non-relativistic form in Eq. (42). μ is, of course, in both equations, constant in space.

7 Model of Confined Atoms in Arbitrary Static Electric Field

Amovilli, March and Pfalzner [41] have studied the inhomogeneous electron cloud in atomic ions 'confined' in hot plasmas and subjected to high static electric fields because of a body of experimental data on multiphoton ionization [42, 43].

In the above study, the aim was first to present a soluble model for a confined assembly of independent electrons subjected to a static electric field of arbitrary strength F. These workers achieved the confinement by imposing a harmonic force in addition to the electric field. They aimed, secondly, to relate their results to atomic ions in hot, non-degenerate plasma.

Below, therefore, the solution of the Bloch equation in Eq. (6) for the canonical density matrix $C(\mathbf{r}, \mathbf{r}_0, \beta, F, \omega)$ for independent electrons in a constant electric field of strength F, with harmonic restoring force corresponding to an oscillator angular frequency ω, will be presented. In Sect. 7.1 below, the electric field is taken as the z axis. Then this solution can readily be generalized to include harmonic restoring forces also in the x and y directions.

Sections 7.3–7.5 are then concerned with relating the above model to atomic ions in a hot, non-degenerate plasma in an external electric field. The first step is to add an 'atomic-like' potential energy $V(\mathbf{r})$ to the model. Strictly, $V(\mathbf{r})$ should be calculated self-consistently as a function of β, F and the plasma density. While this has not been achieved numerically at the time of writing, a model potential $V(\mathbf{r})$ is incorporated into the treatment of Sect. 7.3 by means of the semiclassical Thomas–Fermi approximation. The second step taken by Amovilli et al. [41] is to connect the strength of the harmonic potential with the plasma density (Sect. 7.4).

Then in Sect. 7.5, some typical numerical examples taken from the work of Amovilli et al. will be presented for realistic values of field, temperature and plasma density.

7.1 Confining Harmonic Force only along Field Direction

Motivated by the form of Eq. (21), one can write, since the motion in x and y directions is unaffected:

$$C(\mathbf{r}, \mathbf{r}_0, \beta, F, \omega) = \frac{1}{2\pi\beta} \exp\left(-\frac{(x - x_0)^2}{2\beta} - \frac{(y - y_0)^2}{2\beta} \right) C_z(z, z_0, \beta, F, \omega). \tag{67}$$

Evidently the differential equation for C_z on the right of Eq. (67) can be readily obtained. The point to be stressed is that the potential terms in the Hamiltonian H can be rearranged as

$$\frac{1}{2} m\omega^2 z^2 - eFz = \frac{1}{2} m\omega^2 \left[z - \frac{eF}{m\omega^2} \right]^2 - \frac{e^2F^2}{2m\omega^2}. \tag{68}$$

Thus, one has to deal with a harmonic oscillator with a shift of origin proportional to electric field. Using the study of Stephen and Zalewski [44], following the earlier work of Sondheimer and Wilson [13] one can show after some calculation that the form of C_z on the right of Eq. (67) is [41]

$$\begin{aligned} C_z(z, z_0, \beta, F, \omega) = {} & \left[\frac{\omega}{2\pi \sinh(\beta\omega)} \right]^{1/2} \\ & \times \exp\left(-\frac{\omega}{4} \tanh\left(\frac{\beta\omega}{2}\right) \left(z + z_0 - \frac{2F}{\omega^2} \right)^2 \right. \\ & \left. -\frac{\omega}{4} \coth\left(\frac{\beta\omega}{2}\right) (z - z_0)^2 \right) \times \exp\left(\frac{\beta F^2}{2\omega^2} \right). \end{aligned} \tag{69}$$

If the limit $\omega \to 0$ is taken in Eq. (69) and the result is inserted into Eq. (67) one then obtains Eq. (24) which was given earlier by Harris and Cina [16]. It can be directly verified by insertion of Eqs. (69) and (24) into the appropriate forms of the Bloch equation at Eq. (6) that they are solutions, and the limit $\beta \to 0$ in each is readily shown to satisfy Eq. (7).

Plots will be given below, in Sect. 7.5, of C_z in Eq. (69) on the diagonal $z_0 = z$ for realistic values of β, F and ω, the last of these being connected with the plasma density in Sect. 7.5 below. However, before turning to that, the interest in atomic ions confined in plasma means that the electronic motion should be confined also in the x and y directions. Since there is axial symmetry around the field direction, this necessitates the introduction of only one further force constant or equivalently a further frequency which will be denoted by ω_1.

7.2 Additional Harmonic Force Confinement in x and y Directions

Using the results of Stephen and Zalewski [44] it is a straightforward matter to introduce the new potential energy contribution $\frac{1}{2} m\omega_1^2 (x^2 + y^2)$ into the free-particle terms in x and y on the right-hand side of Eq. (67). Then the new form of Eq. (67) reads [41]

$$C(\mathbf{r}, \mathbf{r}_0, \beta, F, \omega, \omega_1) = \left[\frac{\omega_1}{2\pi \sinh(\beta\omega_1)}\right] \times \exp\left(-\frac{\omega_1}{4}\tanh\left(\frac{\beta\omega_1}{2}\right)(x+x_0)^2 - \frac{\omega_1}{4}\coth\left(\frac{\beta\omega_1}{2}\right)(x-x_0)^2\right) \times \exp\left(-\frac{\omega_1}{4}\tanh\left(\frac{\beta\omega_1}{2}\right)(y+y_0)^2 - \frac{\omega_1}{4}\coth\left(\frac{\beta\omega_1}{2}\right)(y-y_0)^2\right) C_z(z, z_0, \beta, F, \omega). \quad (70)$$

Again representative plots of Eq. (70) on the diagonal $\mathbf{r} = \mathbf{r}_0$ will be presented in Sect. 7.5.

This is the point to turn to the way one now introduces the atomic ion, which is modelled through a suitable one-body potential $V(\mathbf{r})$. In general, self-consistent determination of $V(\mathbf{r})$ in a plasma will lead to the potential depending not only on $\mathbf{r}$ but also on temperature and electric field. Though such self-consistency is not attempted here, some discussion will be given in Sect. 7.5 of the regime where the field dependence of $V(\mathbf{r})$ might be unimportant.

7.3 Introduction of Model Potential Energy $V(\mathbf{r})$ Representing Atomic Ion

Let us now turn to the problem switching on a model potential $V(\mathbf{r})$ to the Hamiltonian used above. Denoting the canonical density matrix calculated there by $C^{(0)} \equiv C(V=0)$, the simplest approximation is to follow the ideas of the Thomas–Fermi (TF) method. Then, with slowly varying $V(\mathbf{r})$ for which the assumptions of this approximation are valid, one can return to the definition at Eq. (2.2), and simply move all eigenvalues ε_i by the same (almost constant—) amount $V(\mathbf{r})$, the wavefunctions $\psi_i(\mathbf{r})$ being unaffected to the same order of approximation. Hence one can write for the diagonal form of the canonical density matrix

$$C(\mathbf{r}, \beta) = C^{(0)} \exp(-\beta V(\mathbf{r})). \quad (71)$$

It is relevant to the discussion of Sect. 4 to note that if V were simply the electric field term $-eFz$ in Eq. (22) and this was switched on the free particle form at Eq. (21), then the additional factor multiplying $C^{(0)}$ would be $\exp(\beta Fz)$. This is precisely the factor present in the diagonal form of Eq. (24). However, potentials $V(\mathbf{r})$ in atomic ions evidently have Coulomb singularities at nuclei, so that

Eq. (71) is a less favourable approximation in this case than for the linear potential $-eFz$.

7.3.1 Transcending Thomas–Fermi Approximation

As proposed by Hilton et al. [45], one can contemplate generalizing the form at Eq. (8) by writing

$$C(\mathbf{r}, \beta) = C^{(0)} \exp[-\beta U(\mathbf{r}, \beta)] \tag{72}$$

where the so-called effective potential U now becomes a function of β, *even if the model potential* $V(\mathbf{r})$ is chosen to be independent of temperature. Hilton et al. propose then to calculate U to first-order only in V. To illustrate their results, if $C^{(0)}$ is replaced by the free-particle limit in zero field, then the first order term of U, say U_1, can be written explicitly in the non local form

$$U_1(\mathbf{r}, \beta) = \int d\mathbf{r}_1 \, V(\mathbf{r}_1) G_0(\mathbf{r}, \mathbf{r}_1, \beta) \tag{73}$$

where

$$G_0(\mathbf{r}, \mathbf{r}_1, \beta) = \frac{1}{\pi\beta|\mathbf{r}_1 - \mathbf{r}|} \exp\left[-\frac{2(\mathbf{r}_1 - \mathbf{r})^2}{\beta}\right]. \tag{74}$$

In Appendix A, Eq. (73) is generalized to apply to switching $V(\mathbf{r})$ on to the model problem of Sect. 7.1. If one restricts oneself here to Eq. (73), Hilton et al. plot $U_1(\mathbf{r}, \beta)$ for various cases: the Coulomb singularity at $\mathbf{r} = 0$ is removed by the non-local form at Eq. (73) for any finite β.

7.4 Connection of Harmonic Confining Force Constants and Frequencies with Plasma Density

One application of the above described model is to the modelling of plasmas. In the statistical description of dense plasmas it is a common method to estimate the radius of the cell occupied per atom by dividing the volume by the number of particles

$$r = \left(\frac{4\pi n_i}{3}\right)^{-1/3}, \tag{75}$$

where n_i is the ion number density. In the case of harmonic confining forces, the radius of this cell can be set equal to the wavelength of the harmonic force. This boundary is, unlike that in the Thomas–Fermi model, a smooth well because of the harmonic potential. The advantage of this boundary is that the electrons are not totally fixed in their cell but some tunneling is allowed as well. So the connection between the frequency of the harmonic force and the plasma density is given by

$$\omega = \pi\left(\frac{4\pi n_i}{3}\right)^{1/3}. \tag{76}$$

This definition of the force constant will be employed below in some illustrative examples. The division of the volume of the plasma into these small cells is best applicable in the case of dense plasmas. The model described above is restricted in the density range because of the assumption of a non-degenerate plasma, but using Fermi–Dirac statistics instead of Maxwell–Boltzmann the range of applicability of this approach could be widened to embrace very high densities ($\sim 10^{23}$ particles per cc).

7.5 Simple Illustrative Examples

Here, some numerical examples will be presented from Amovilli et al. [41]. As far as possible, bearing in mind the limitations of the model, the examples are designed for conditions which can be achieved in laboratory experiments. However only non-degenerate plasmas will be considered, this then implying the constraint that the ionic number density n_i satisfies

$$n_i \ll 1.4 \cdot 10^{23}\,\frac{1}{Z}\left(\frac{k_B T}{10\mathrm{eV}}\right)^{3/2} \tag{77}$$

with Ze the charge of an ion. This delineates the region of classical plasmas. The most recent experiments have considered higher densities, where the effect of the degeneracy of the electrons becomes important. However in the context of multiphoton ionization relatively low density plasmas are normally investigated. The presently achievable laser flux is about 10^{18} Watt/m^2. The connection between the laser flux and the electric field is given by

$$I = cE_{\max}^2/8\pi. \tag{78}$$

In the above calculation we have assumed a static electric field. Brewczyk and Gajda [43] pointed out under what conditions this is a reasonable assumption.

Figure 1 shows the temperature, density and electric field dependence as represented by Eq. (69). The behaviour of C_z in the density and temperature region on which we focus is dominated by the temperature dependence which is $\sim T^{1/2}$. The E-field and density dependence is stronger the lower the temperature. The plot is for $z = z_0$ and z substantially less than E/ω^2. Figure 2 similarly shows how the x and y terms contribute to C, namely the ratio C/C_z from Eq. (67), depends on temperature and density. Here the temperature dependence is $\sim T$ and again the density dependence is the stronger the lower the temperature.

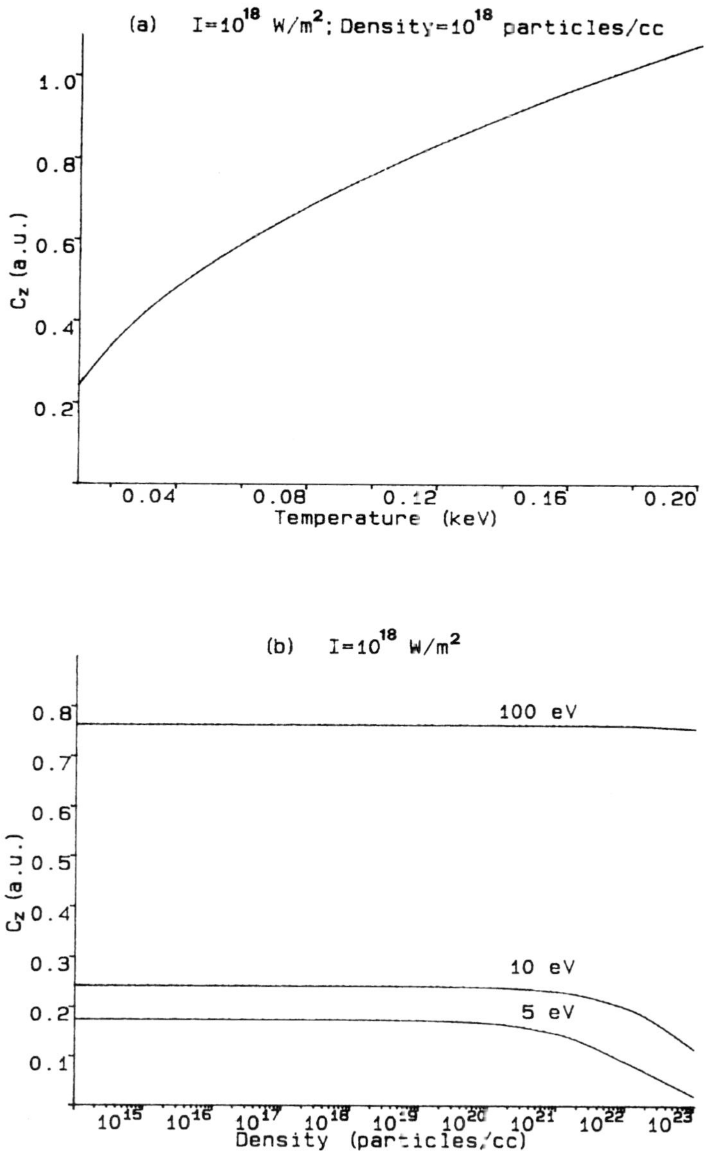

Fig. 1a, b. Variation of non-degenerate density C_z in Eq (7): **a** with temperature at fixed density 10^{18} particles/cc; **b** with density at the temperatures corresponding to $k_B T = 5$, 10 and 100 eV. In the calculation $z = z_0$ and $z \ll E/\omega^2$. The equivalent laser flux is 10^{18} Watt/m^2. After Amovilli et al. [41]

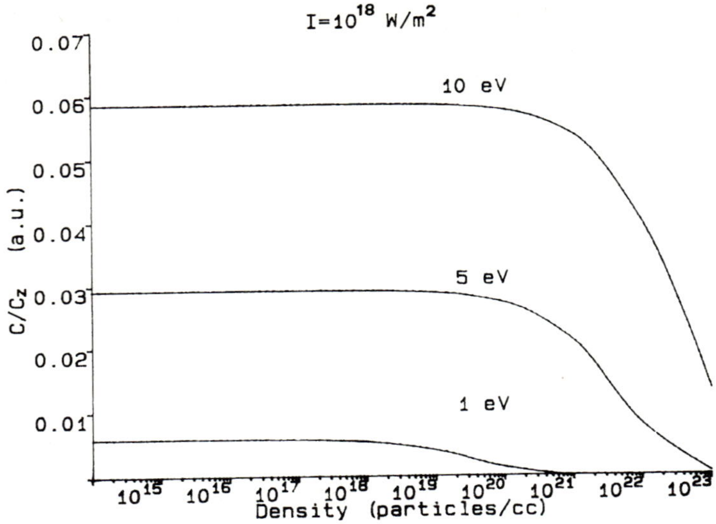

Fig. 2. Same as Fig. 1b but now for the x and y contribute to the non-degenerate density C from Eq. (9), and with different temperatures corresponding to $k_B T = 1, 5$ and 10 eV. It should be noted for $k_B T = 1$ eV, C begins to decrease at densities of $\sim 10^{18}$–10^{19} particle/cc. (After Amovilli et al. [41])

In short, closed forms have been obtained for the canonical density matrix C for electrons moving in a static electric field E, and confined by a harmonic restoring force. Model potentials $V(\mathbf{r})$ have then 'switched on' to this above canonical density matrix via the TF approximation at Eq. (71).

It would be of interest to apply the method of March and Murray [12] to convert C, the electron density for non-degenerate electrons, into results applicable to intermediate degeneracy governed by Fermi–Dirac statistics. Unfortunately, without switching on the model potential $V(\mathbf{r})$, this is already difficult to handle by purely analytically methods, as can be seen from the case of complete degeneracy for the harmonic oscillator alone. No doubt, numerical procedures will eventually enable present results to be transformed according to the route established in [12].

The same situation applies when one attempts to remove the TF approximation underlying Eq. (71). With $C^{(0)}$ instead of the free-particle C_0, the generalization of the Green function G_0 in Eq. (73) is hard to effect analytically. Numerical presentation will be difficult, because of the large number of variables involved.

Nevertheless, it seems likely that the model treatment of atomic ions in hot, non-degenerate plasmas presented in this work, is well worth further study, the intermediate Fermi–Dirac degeneracy being of obvious importance. Under these conditions, an appropriate starting point to introduce the potential would be the elevated temperature Thomas–Fermi theory [46].

8 Time-Dependent Uniform Electric Field Propagator

Throughout this section, the canonical density matrix and the Feynman propagator can be used interchangeably, the transformation $\beta = it$ taking C into the propagator K, with t the time. While most frequently we shall use the coordinate representation $\mathbf{r}$ and $\mathbf{r}'$, it will be convenient in this section to work in $\mathbf{k}$ or momentum representation, by taking a double Fourier transform with respect to $\mathbf{r}$ and $\mathbf{r}'$.

Let us first then effect the generalization of free electrons in a static uniform electric field F to treat a uniform time-varying electric field. The Feynman propagator for this case has been discussed, for example, by Fallieros and Friar [47]. These workers give the propagator $K(\mathbf{k}, \mathbf{k}', t)$ in $\mathbf{k}$ space as

$$K(\mathbf{k}, \mathbf{k}', t) = (2\pi)^3 \delta\left(\mathbf{k} - \mathbf{k}' - eF\frac{\sin \omega t}{\hbar\omega}\right) \exp(i\chi_\omega(k, t)). \tag{79}$$

Here the phase function χ_ω can then be obtained explicitly. For zero field $F = 0$, it is readily obtained from an earlier section as $-\varepsilon_{\mathbf{k}} t$ where $\varepsilon_{\mathbf{k}}$ is the free-electron energy $\hbar^2 k^2/2m$. When the field is switched on the phase becomes

$$\chi_\omega(k, t) = -\frac{\hbar^2 k^2}{2m} t - e\mathbf{F} \cdot \mathbf{k} \frac{(1 - \cos \omega t)}{m\omega^2} - e^2 F^2 \frac{(2\omega t - \sin(2\omega t))}{8m\hbar\omega^3}. \tag{80}$$

This work of Fallieros and Friar also deals with the momentum probability amplitude: we refer the reader to their study. However, for the moment, consider the phase in the limit of a static field by allowing the angular frequency ω to tend to zero. Then the phase χ_0 becomes $-\varepsilon_{\mathbf{k}} t - e\mathbf{F} \cdot \mathbf{k}\, t^2/m - \frac{1}{6} e^2 F^2 t^3/m\hbar$. Taking the double $\mathbf{r}$ space Fourier transform of this limit, one naturally recovers the static field case discussed in Sect. 4.2.

9 Molecular Ions in Intense Magnetic Fields

Though this section is mainly about atomic ions, we shall add a little on molecular ions, beyond semiclassical theory. It is natural then to start with H_2^+. Though a good deal of work has been done, we shall comment only on the recent study of Amovilli and March [48] in the present context.

9.1 H_2^+ Ion: Field along Internuclear Axis

Below, we shall write an expression for the ground-state energy of the H_2^+ ion with a magnetic field of strength B applied along the internuclear axis. We denote the ground-state energy by $E^+(R; B)$ for internuclear separation R. It will be shown that E can be expressed explicitly in terms of density, potential and current density (see below and also Appendix B).

Employing units in which the electronic charge is -1, mass 1 and $\hbar = 1$, the Hamiltonian H takes the form, with V written for potential energy and with neglect of the interaction between spin magnetic moment and field

$$H = \tfrac{1}{2}(-\mathrm{i}\nabla + \mathbf{A}/c)^2 + V = H_{\mathrm{OB}} + V, \tag{81}$$

the vector potential being taken for the constant magnetic field $\mathbf{B}$ as

$$\mathbf{A} = \tfrac{1}{2}\mathbf{B}\times\mathbf{r}. \tag{82}$$

The wave function ψ for the ground state will now be written as the density amplitude $\{\rho(\mathbf{r})\}^{1/2}$ already introduced in Sect. 6.2 times a phase factor $\exp(i\theta)$. It is useful for what is to follow to introduce the current density $\mathbf{j}$ at this stage (see also Appendix B). This is given, in the presence of the magnetic field, by

$$\mathbf{j} = \frac{\mathrm{i}}{2}(\psi\nabla\psi^* - \psi^*\nabla\psi) - \frac{\mathbf{A}}{c}\psi\psi^* = \rho\nabla\theta + \frac{\mathbf{A}}{c}\rho. \tag{83}$$

Then one obtains for the ground-state energy $E^+(R, B)$:

$$E^+(R, B) = \frac{H\psi}{\psi} = \frac{(\nabla\rho)^2}{8\rho^2} - \frac{\nabla^2\rho}{4\rho} + \frac{\mathbf{j}^2}{2\rho^2} + \frac{\mathrm{i}\nabla\cdot\mathbf{j}}{2} + V, \tag{84}$$

this equation applying at any point $\mathbf{r}$. For stationary wave functions, and hence time-independent electron density $\rho(\mathbf{r}, R, B)$, it follows from the equation of continuity that

$$\operatorname{div}\mathbf{j} = 0 \tag{85}$$

and hence

$$E^+(R, B) = \frac{(\nabla\rho)^2}{8\rho^2} - \frac{\nabla^2\rho}{4\rho} + \frac{j^2}{2\rho^2} + V, \tag{86}$$

which is the result of Amovilli and March [48]. Current density considerations are developed further in Appendix B.

9.2 Molecular Anions: Terrestrial and in Intense Magnetic Fields

To finish this brief section, we draw attention, at the time of writing, to interest in molecular anions in chemistry. By way of the briefest introduction, the TF heavy, non-relativistic, neutral atom in zero magnetic field has zero chemical potential. As a consequence, one does not have a negative atomic ion in the limit of large atomic number Z. In other words, a nucleus of charge Ze can bind, at most, Z electrons in the limit Z tends to infinity in Schrödinger wave mechanics.

However, the work of Lieb et al. [36], discussed in Sect. 6.3, shows that this situation will be completely changed in hyperstrong magnetic fields: e.g. in the field of a neutron star. There, atomic 'needles' can bind only just less than $2Z$ electrons, and multiply-charged negative ions will readily bind. For molecular anions, the recent work of Freeman and March [39] has discussed what molecular dianions presently exist terrestrially, and has pointed out the possible interest in studying negative molecular ions at the highest magnetic fields available in the laboratory. Of course, these are hardly 'intense' fields as yet compared with their galactic counterparts!

10 Summary and Future Directions

For intense magnetic fields, and for heavy positive ions, the scaling properties of the non-relativistic ground-state energy $E(N, Z, B)$ have been established. Numerical self-consistent calculations are now available from Thomas–Fermi theory for $T = 0$ (Fermi–Dirac completely degenerate limit) for fields of the order of 10^{13} gauss. For 'hyperstrong' fields, the Boson equation for the density amplitude $\{\rho(\mathbf{r})\}^{1/2}$ with a suitable Pauli potential has been solved analytically by Lieb et al. [36]. The atomic 'needles' that are then formed can bind many extra electrons (up to $N = 2Z$). Terrestrial chemistry is thus completely changed in magnetic fields of neutron stars. In particular, homonuclear dissociation energies will be of the order of atomic binding energies.

Turning to electric fields and classical Maxwell–Boltzmann statistics, soluble analytical models now exist which allow calculations of non-degenerate electron densities as a function of thermodynamic state in intense electric fields (low density: high temperature). Semiclassical methods are available for 'switching on' atomic potentials to models studied presently, though numerical results are not yet available here.

As to future directions, the problem of the canonical density matrix, or equivalently the Feynman propagator, for hydrogen-like atoms in intense external fields remain an unsolved problem of major interest. Not unrelated, differential equations for the diagonal element of the canonical density matrix, the important Slater sum, are going to be worthy of further research, some progress having already been made in (a) intense electric fields and (b) in central field problems. Finally, further analytical work on semiclassical time-dependent theory seems of considerable interest for the future.

Appendix A Effective Potential and Green Function

The approach of Hilton et al. can be generalized for any reference Hamiltonian for which the solution of the relevant Bloch equation is known. For a given Hamiltonian

$$H_0 = -\tfrac{1}{2}\nabla^2 + V_0(\mathbf{r}) \tag{A1}$$

it is possible to find the solution of the Bloch equation for a perturbed Hamiltonian

$$H = H_0 + V(\mathbf{r}). \tag{A2}$$

The procedure is to write

$$C(\mathbf{r}, \mathbf{r}_0, \beta) = C_0(\mathbf{r}, \mathbf{r}_0, \beta)\exp[-\beta U(\mathbf{r}, \mathbf{r}_0, \beta)] \tag{A3}$$

where $C_0(\mathbf{r}, \mathbf{r}_0, \beta)$ satisfies the equation

$$\hat{H}_0 C_0 = -\frac{\partial C_0}{\partial \beta}. \tag{A4}$$

The effective potential matrix $U(\mathbf{r}, \mathbf{r}_0, \beta)$ satisfies the equation

$$\left[1 + \beta\frac{\partial}{\partial\beta} - \beta\frac{\nabla C_0}{C_0}\cdot\nabla - \frac{\beta}{2}\nabla^2\right]U = V - \tfrac{1}{2}\beta^2|\nabla U|^2. \tag{A5}$$

As in the approach of Hilton et al. this differential equation in U can be transformed into an integral equation by using the Green function of the left-hand side operator in Eq. (A5). This Green function maintains the same form as for the solution of Hilton et al. for the perturbed non-interacting free-electron system, namely

$$G(\mathbf{r}, \mathbf{r}_0, \mathbf{r}_1, \beta, \beta_1) = \frac{C_0(\mathbf{r}, \mathbf{r}_1, \beta - \beta_1)C_0(\mathbf{r}_1, \mathbf{r}_0, \beta_1)}{\beta C_0(\mathbf{r}, \mathbf{r}_0, \beta)}\theta(\beta - \beta_1) \tag{A6}$$

but now C_0 is the solution of the Bloch equation at Eq. (6) for the reference Hamiltonian at Eq. (A1). The corresponding integral to Eq. (A5) is then

$$U(\mathbf{r}, \mathbf{r}_0, \beta) = \int d\mathbf{r}_1 \int_0^\beta d\beta_1\, G(\mathbf{r}, \mathbf{r}_0, \mathbf{r}_1, \beta, \beta_1)\left[V(\mathbf{r}_1) - \frac{\beta^2}{2}|\nabla U|^2\right]. \tag{A7}$$

When it is possible to neglect the term $(\beta^2/2)|\nabla U|^2$, Eq. (A7) gives a direct route for calculating the “effective potential” U. This linear response treatment can be applied under the following conditions: (i) U small and much more slowly varying with $\mathbf{r}$ than $V(\mathbf{r})$, especially in presence of Coulomb singularities, (ii) $|\nabla U|^2$ small, (iii) β small. Using Eq. (A4) for C_0 in (A6) and (A7) it is not possible to give an analytical expression for U even in the linear response approximation and numerical procedures are required. When it is possible to make the assumption $\beta\omega < 0.5$ the β-convolution in Eq. (A7) can be computed by approximating the hyperbolic functions of C_0 by the lowest order powers in $\beta\omega$. In the linear response approximation in V and in this high temperature regime, Eq. (A7) takes the form, for the diagonal elements,

$$U(\mathbf{r}, \beta) = \int d\mathbf{r}_1 V(\mathbf{r}_1) G(\mathbf{r}, \mathbf{r}_1, \beta) \tag{A8}$$

where [41]

$$G(\mathbf{r}, \mathbf{r}_1, \beta) = \frac{1}{\pi\beta|\mathbf{r}_1 - \mathbf{r}|} \exp\left\{ -\frac{2(\mathbf{r}_1 - \mathbf{r})^2}{\beta} + \frac{\beta F}{2}(z_1 - z) - \frac{\beta\omega^2}{8}[(\mathbf{r}_1 + \mathbf{r})^2 - 4r^2] - \frac{\beta\omega^2}{24}(\mathbf{r}_1 - \mathbf{r})^2 \right\}. \tag{A9}$$

This reduces to the free-electron Green function when $F = 0$ and $\omega = 0$.

Appendix B Considerations on Current Density

The motion of electrons in a magnetic field in a situation in which inhomogeneity of some kind exists remains of considerable interest at the time of writing. Therefore, in this Appendix, we shall first summarize some results of Freeman and March [49] for the current density in a simple model of independent harmonically confined electrons in a constant magnetic field. Then we shall go on to discuss the semiclassical theory of current density in atoms.

Freeman and March [49] exploit the progress in calculating the canonical density matrix of independent electrons moving under the combined effect of a harmonically confining force and a constant magnetic field B along the z axis [50]. The vector potential $\mathbf{A}$ in the resulting Hamiltonian H given by

$$H = \frac{(\mathbf{p} - e\mathbf{A}/c)^2}{2m} + \tfrac{1}{2}k(x^2 + y^2) \tag{B1}$$

is chosen in the so-called symmetric gauge in which

$$\mathbf{A} = (-\tfrac{1}{2}By, \tfrac{1}{2}Bx, 0). \tag{B2}$$

The structure of the canonical density matrix is then

$$C(\mathbf{r}_0, \mathbf{r}, \beta) = f(\beta)\exp\{-i(x_0 y - y_0 x)\phi(\beta) - [(x - x_0)^2 + (y - y_0)^2]g(\beta) - [(x + x_0)^2 + (y + y_0)^2]h(\beta)\}. \tag{B3}$$

By substituting this form in the Bloch equation at Eq. (6), March and Tosi [50] determined the functions $f(\beta)$, $g(\beta)$ and $h(\beta)$ explicitly.

The important Slater sum $S(\mathbf{r}, \beta)$ is therefore from Eq. (B3):

$$S(\mathbf{r}, \beta) = f(\beta) \exp[-4(x^2 + y^2)h(\beta)]. \tag{B4}$$

Two points associated with Eq. (B4) are noteworthy. First, the function $\phi(\beta)$, which appears in the phase of the density matrix in Eq. (B3) does not enter the Slater sum at Eq. (B4). Secondly, $S(\mathbf{r}, \beta)$ falls off in Gaussian fashion with the distance from the origin of the confining potential. However, since $h(\beta)$ has the form

$$h(\beta) = \frac{m\omega b}{4\hbar} \coth(b\alpha) - \frac{\cosh \alpha}{\sinh(b\alpha)} \tag{B5}$$

where $\omega = eB/2mc$ is the Larmor angular speed, $\alpha = \hbar\omega\beta$, while

$$b = \left\{1 + \frac{k}{m\omega^2}\right\}^{1/2} \tag{B6}$$

it follows from Eqs. (B4) and (B5) that the inhomogeneity in the Slater sum $S(\mathbf{r}, \beta)$ is solely due to the confining force, since when k tends to zero, b tends to unity from Eq. (B6) and hence $h(\beta)$ tends to zero from Eq. (B5). Naturally, this reflects the translational invariance already present in the Sondheimer and Wilson [13] result for free electrons moving in a constant magnetic field.

Let us turn then to the calculation of the current density $\mathbf{J}(\mathbf{r}, \beta)$, following the work of Amovilli and March [51]: see also below. One can write $\mathbf{J}(\mathbf{r}, \beta)$ in terms of the canonical density and its diagonal-the Slater sum, as (in suitable units)

$$\mathbf{J}(\mathbf{r}, \beta) = + \ -\frac{i}{2}(\nabla_0 - \nabla)C(\mathbf{r}_0, \mathbf{r}, \beta)|_{r_0=\mathbf{r}} - \mathbf{B} \times \mathbf{r}\, S(\mathbf{r}, \beta). \tag{B7}$$

Using the result at Eq. (B3), one notes that $\mathbf{J}$ has the form

$$\mathbf{J} = J_x\mathbf{i} + J_y\mathbf{j} \tag{B8}$$

with $\mathbf{i}$ and $\mathbf{j}$ the Cartesian unit vectors in the x and y directions respectively. One then readily finds from Eqs. (B3) and (B7)

$$\mathbf{J}_x = -y[\phi(\beta) - \phi(\beta, k=0)]\, S(\mathbf{r}, \beta) \tag{B9}$$

with an analogous expression for J_y. Here, from March and Tosi [50], the phase function $\phi(\beta)$ is given by

$$\phi(\beta) = \frac{m\omega b}{\hbar} \frac{\sinh \alpha}{\sinh(b\alpha)}. \tag{B10}$$

With these results for a simple model as background we now summarize some semiclassical results for the current density in atoms.

Atomic Current Density: Limits of Intense and Weak Fields

Following the work of Hilton, March and Curtis [45] and the later study of March and Stoddart [52], one can write, in the presence of an atomic potential $V(\mathbf{r})$

$$C(\mathbf{r}, \mathbf{r}_0, \beta) = C_0(\mathbf{r}, \mathbf{r}_0, \beta) \exp(-U(\mathbf{r}, \mathbf{r}_0, \beta)). \tag{B11}$$

From the Bloch equation, one can then obtain a nonlinear equation for the so-called effective potential matrix $U(\mathbf{r}, \mathbf{r}_0, \beta)$. The central approximation in what is to follow is to linearize this equation, which is the same as replacing U by U_1, where U_1 is now only of $O(V)$. The result for U_1 from the work of March and Stoddart is

$$U_1(\mathbf{r}, \mathbf{r}_0, \beta) = \int d\mathbf{r}_1 \int_0^\beta d\beta' \frac{C_0(\mathbf{r}, \mathbf{r}_1, \beta - \beta') C_0(\mathbf{r}_1, \mathbf{r}_0, \beta')}{\beta C_0(\mathbf{r}, \mathbf{r}_0, \beta)} V(\mathbf{r}_1). \tag{B12}$$

Given this form, the current follows by inserting $C_1 = C_0 \exp(-\beta \mathrm{U}_1)$ for C in Eq. (B7) to yield after a short calculation

$$\mathbf{J}(\mathbf{r}, \beta) = \frac{i\beta}{2} C_1(\mathbf{r}, \mathbf{r}_0, \beta)(\vec{\nabla} - \vec{\nabla}_0) U_1(\mathbf{r}, \mathbf{r}_0, \beta)|_{\mathbf{r}_0 = \mathbf{r}}. \tag{B13}$$

It is convenient now to define the matrix, following Amovilli and March [51],

$$\Gamma(\beta, B) = \begin{pmatrix} -\frac{B}{2}\coth(\beta B) & 0 & 0 \\ 0 & \frac{B}{2}\coth(\beta B) & 0 \\ 0 & 0 & \frac{1}{2\beta} \end{pmatrix}, \tag{B14}$$

and with the notation $C_0(\mathbf{r}, \mathbf{r}_0, \beta) \equiv K(\beta, B)$, one finds after some calculation that the quantity $\mathbf{J}/C_1$ from Eq. (B13) has the form

$$\frac{\mathbf{J}(\mathbf{r}, \beta)}{C_1(\mathbf{r}, \mathbf{r}, \beta)} = \int d\mathbf{r}_1 V(\mathbf{r}_1) \mathbf{G}_\beta(\mathbf{r}_1 - \mathbf{r}). \tag{B15}$$

This is evidently a nonlocal relation between $\mathbf{J}$ and V, the kernel $\mathbf{G}_\beta$ being given explicitly by [51]

$$\mathbf{G}_\beta(\mathbf{r}_1 - \mathbf{r}) = -\mathbf{B} \times (\mathbf{r}_1 - \mathbf{r}) \int_0^\beta d\beta' \frac{K(\beta - \beta', B) K(\beta', B)}{K(\beta, B)}$$
$$\times \exp[-(\mathbf{r}_1 - \mathbf{r}) \cdot (\Gamma(\beta - \beta') + \Gamma(\beta')) \cdot (\mathbf{r}_1 - \mathbf{r})]. \tag{B16}$$

Strong Magnetic Field Limit

Taking first the high field limit one readily finds from Eq. (B16)

$$K(\beta, B) = \frac{B}{(2\beta)^{1/2}\pi^{3/2}} \exp(-\beta B) \tag{B17}$$

$$\Gamma(\beta, B) = \begin{pmatrix} -\frac{B}{2} & 0 & 0 \\ 0 & \frac{B}{2} & 0 \\ 0 & 0 & \frac{1}{2\beta} \end{pmatrix} \tag{B18}$$

and after a certain amount of manipulation, with $\mathbf{R} = \mathbf{r}_1 - \mathbf{r}$,

$$\mathbf{G}_\beta(\mathbf{R}) = -\mathbf{B} \times \mathbf{R} \frac{\beta^{1/2} B}{(2\pi)^{1/2}} \exp\left[-\frac{B}{2}(X^2 + Y^2)\right] \operatorname{erfc}\left[\frac{2|Z|}{(2\beta)^{1/2}}\right], \tag{B19}$$

where

$$\operatorname{erfc}(x) = \frac{2}{\pi^{1/2}} \int_x^\infty \exp(-u^2)\, du (x \geqslant 0). \tag{B20}$$

Weak Field limit

Here, K and Γ are readily calculated in the $B \to 0$ limit and one obtains

$$\mathbf{G}_\beta(\mathbf{R}) = -\mathbf{B} \times \mathbf{R} \frac{1}{\pi R} \exp\left[-\frac{2R^2}{\beta}\right]. \tag{B21}$$

Performing a gradient expansion of the potential in Eq. (B15) as

$$V(\mathbf{r}_1) = V(\mathbf{r}) + (\nabla V)_\mathbf{r} \cdot (\mathbf{r}_1 - \mathbf{r}), \tag{B22}$$

one finds

$$\frac{\mathbf{J}(\mathbf{r}, \beta)}{C_1(\mathbf{r}, \mathbf{r}, \beta)} = -\frac{\beta^2}{6} \mathbf{B} \times \nabla V. \tag{B23}$$

This reduces to Eq. (18) of Harris and Cina [16] when β is replaced by *it*, with t the time, and one approximates also C_1 by the lowest order gradient expansion as $C_0 \exp(\beta V(r))$. This procedure then satisfies the equation of continuity $\nabla \cdot \mathbf{J} = 0$ at this order of approximation.

Returning to the strong field limit, the same gradient expansion at Eq. (B22) would lead to the result $\mathbf{J}/C_1 = -(2\beta/B)\mathbf{B} \times \nabla V$, showing then that the strength of the magnetic field cancels out. However, the kernel $\mathbf{G}_\beta$ has two equal and opposite peaks away from the diagonal, and care may well be needed in evaluating Eq. (B15) for given V, by means of Eq. (B22). Furthermore, in this high field regime, it will no doubt eventually be necessary to calculate $V(\mathbf{r})$ self-consistently, and then the field dependence will clearly enter, with cylindrical symmetry around the field direction, rather than spherical symmetry when $B = 0$.

Equation (B15) is the main result of this Appendix. The way to pass from this equation to the weak field limit given by Harris and Cina [16] has been noted above. The derivation [51] of the current in Eq. (B21) complements the discussion of charge density given by Li and Percus [53].

Acknowledgments. It is a pleasure to acknowledge many valuable discussions and much invaluable collaboration with my friends and colleagues C. Amovilli, G.R. Freeman, A. Holas and S. Pfalzner. Also, I am indebted to Dr. P. Schmidt for much encouragement, and for partial financial support for work in the area, which he initiated, through the Office of Naval Research, USA. Finally, the opportunity to write a substantial part of this paper arose from a visit to ITAMP in early 1995. I wish to thank especially Professors A. Dalgarno and M. Gavrila for much fruitful interaction on the general area of the present work and for kind hospitality. The work was partly supported by the National Science Foundation through a grant for ITAMP at Harvard University and the Smithsonian Astrophysical Observatory.

References

1. Garstang RH (1982) J Physique Coll C2 Suppl 11: 4319
2. Trümper J, Pietsch W, Reppin C, Voges W, Staubert R, Kandziorra E (1978) Astrophys J Lett 239: L107
3. Maurer GS, Johnson WN, Kurfess JD, Strickman MS (1982) Astrophys J 254: 271
4. Kadomtsev BB (1970) Sov Phys JETP 31: 945
5. Mueller RO, Rau ARP, Spruch L (1971) Phys Rev Lett 26: 1136
6. March NH (1992) Electron density theory of atoms and molecules (Academic: New York)
7. Lundqvist S, March NH (1983) Editors: Theory of the inhomogeneous electron gas, Plenum, New York
8. Parr RG, Yang W (1989) Density functional theory of atoms and molecules, Oxford University Press
9. Hill SH, Grout PJ, March NH (1983) J Phys B16: 2301
10. Vallarta MS, Rosen N (1932) Phys Rev 41: 708
11. March NH (1993) Phys Rev A48: 4778
12. March NH, Murray AM (1960) Phys Rev 120: 830
13. Sondheimer EH, Wilson AH (1951) Proc Roy Soc A210: 173
14. Pfalzner S, March NH (1993) J Math Phys 34: 549
15. Jannussis AD (1969) Phys Status Solidi 36: K17
16. Harris RA, Cina J (1983) J Chem Phys 79: 1381
17. March NH, Young WH (1959) Nuclear Physics 12: 237
18. Amovilli C, March NH (1991) Phys Rev A44: 2846
19. Lehmann H, March NH (1994) Phys Chem Liquids 27: 65
20. Amovilli C, March NH (1995) Phys Chem Liquids 30: 135
21. Germann TC, Herschbach DR, Dunn M, Watson DK (1995) Phys Rev Lett 74: 658
22. Herschbach DR, Avery J, Goscinski O (1992) Editors: Dimensional scaling in Chemical Physics: Dordrecht: Kluwer
23. Goodson DZ, Lopez-Cabrera M, Herschbach DR, Morgan JD (1992) J Chem Phys 97: 8481
24. Vainberg VM, Popov VS, Sergeev AV (1990) Sov Phys JETP 71: 470
25. Bender CM, Mlodinow LD, Papanicolau N (1982) Phys Rev A25: 1305

26. Mlodinov LD, Papanicolau N (1980) Annals of Physics 128: 314: ibid (1981) 131: 1
27. Bethe HA, Salpeter EE (1977) Quantum mechanics of one- and two-electron atoms: Plenum-Rosette, New York
28. March NH, Tomishima Y (1979) Phys Rev D19: 449
29. March NH, White RJ (1972) J Phys B5: 466
30. Hill SH, Grout PJ, March NH (1985) J Phys B18: 4665
31. Hill SH, Grout PJ, March NH (1987) J Phys B20: 11
32. March NH, Murray AM (1960) Proc Roy Soc A256: 400
33. Levy M, Görling A (1994) Phil Mag B69: 763
34. Holas A, March NH (1995) Phys Rev A 51: 2040
35. Löwdin PO (1955) Phys Rev 97: 1474, 1490, 1509
36. Lieb EH, Solovej JP, Yngvason J (1992) Phys Rev Lett 69: 749
37. Lehmann H, March NH (1995) Pure and Appl Chem 67: 457
38. Teller E (1962) Rev Mod Phys 34: 627
39. Freeman GR, March NH (1996) J Phys Chem 100: 4331
40. Hill SH, Grout PJ, March NH (1984) J Phys B17: 4819
41. Amovilli C, March NH, Pfalzner S (1991) Phys Chem Liquids 24, 79
42. Szoke A, Rhodes CK (1986) Phys Rev Lett 56: 720
43. Brewczyk M, Gajda M (1989) J Phys B21: 925; Phys Rev A40: 3475
44. Stephen MJ, Zalewski K (1962) Proc Roy Soc A270: 435
45. Hilton D, March NH, Curtis AR (1967) Proc Roy Soc A300: 391
46. Feynman RP, Metropolis N, Teller E (1949) Phys Rev 75: 1561
47. Fallieros S, Friar JL (1982) American J. Phys. 50: 1001
48. Amovilli C, March NH (1990) Chem Phys 146: 207
49. Freeman GR, March NH (1992) Phys Rev A45: 6879
50. March NH, Tosi MP (1985) J Phys A18: L643
51. Amovilli C, March NH (1991) Phys Rev A43: 2528
52. March NH, Stoddart JC (1968) Rep Prog Phys 31: 533
53. Li S, Percus JK (1990) Phys Rev A41: 2344: see also J Chem Phys 93: 4266

Field Induced Chaos and Chaotic Scattering

Harald Friedrich

Physik-Department, Technische Universität München, 85747 Garching, Germany

One-electron atoms subjected to a time-dependent external field provide physically realistic examples of scattering systems with chaotic classical dynamics. Recent work on atoms subjected to a sinusoidal external field or to a periodic sequence of instantaneous kicks is reviewed with the aim of exposing similarities and differences to frequently studied abstract model systems. Particular attention is paid to the fractal structure of the set of trapped unstable trajectories and to the long time behavior of survival probabilities which determine the ionization rates of the atoms. Corresponding results for unperturbed two-electron atoms are discussed.

1 Introduction 98

2 "Generic" Chaotic Scattering 100

3 One-Electron Atom in an Oscillating External Field 104

4 Chaotic Scattering in Two-Electron Atoms 114

5 Summary 120

6 References 122

1 Introduction

One main effect of a strong external field on a simple system such as a one-electron atom is that the integrability of the classical equations of motion of the simple system is destroyed while the quantum mechanical Schrödinger equation can no longer be straightforwardly solved by a separation of variables. It is now generally accepted that most non-integrable classical systems can evolve chaotically, and the study of simple (one-electron) atoms in external fields has become an active area of exciting developments in the study of physically real chaotic systems and of quantum manifestations of classical chaoticity.

Studies of chaos in atomic systems have concentrated on two classes of examples. The first consists of conservative Hamiltonian systems with non-integrable classical dynamics, which are realised by atoms subjected to static external fields. The hydrogen atom in a uniform magnetic field is probably the most widely studied example in this class. The classical transition to chaos, its manifestation in statistical properties of quantum spectra and the role of short unstable periodic classical orbits in shaping the longer ranged structure of the quantum spectra were studied in great detail in the late 1980s and are now well understood [1–3]. More recently, advances have been made in extending the studies to one-electron atoms with non-vanishing quantum defects describing the effect of an ionic core [4, 5], and in understanding the coupling of the internal motion of the hydrogen atom and its centre of mass motion in the presence of the magnetic field [6–9].

The second class of atomic systems studied in the search for manifestations of chaos consists of time-dependent Hamiltonian systems such as one-electron atoms in an oscillating field. The hydrogen atom in a microwave or laser field is the standard physical example and has been a focus of attention since the ionization of highly excited hydrogen atoms by intense microwave fields was first observed by Bayfield and Koch in 1974 [10].

The solution of the classical equations of motion or of the Schrödinger equation for a seemingly simple system such as a one-electron atom in a time-dependent external field is a formidable task. Early investigations devoted to the problem of a hydrogen atom in a linearly polarized microwave field [11–13] revealed that much of the physics of the problem can already be derived from a simple one-dimensional ansatz. Further simplification can be obtained by replacing the sinusoidal dependence of the external field by a periodic sequence of instantaneous kicks; this greatly facilitates the integration of the classical equations of motion. An early success of one-dimensional classical calculations was the prediction of observed threshold field strengths for the microwave ionization of hydrogen atoms from a highly excited initial state with principal quantum number n_0 for microwave frequencies up to the corresponding classical frequency $\omega_0 \approx 1/n_0^3$ ([14]; see also [15]). It is now generally accepted that the onset of ionization in this regime is related to the onset of chaos in the classical system resulting in a diffusive growth of the electron's excitation energy, once the perturbing oscillating field is strong enough to break up the Kolmogorov-Arnold-Moser (KAM) tori of regular near Coulombic motion of the electron characteristic of small field strengths.

Considerable progress has been made in recent years in extending calculations to the more realistic three-dimensional case with linearly or circularly polarized external fields. One important generalization is the allowance of a finite pulse, whereby the external field is no longer strictly periodic but contains an amplitude factor with finite switch-on and switch-off times. Gebarowski and Zakrzewski [16] have studied the classical dynamics of two- and three-dimensional hydrogen atoms in the presence of a circularly polarized external field, Benvenuto et al. [17, 18] and Casati et al. [19] have studied the classical dynamics of the three-dimensional atom in a linearly polarized field approximated by a chain of instantaneous kicks, Wiedemann et al. [20] have compared the dynamics obtained in the kicking approximation to the results obtained using the realistic sinusoidal dependence of the electric field strength; Su et al. [21], Pont and Shakeshaft [22] and Kulander et al. [61] have performed extensive quantum mechanical calculations for the three-dimensional case. The main emphasis of all the papers just mentioned is on the effect of strong field stabilization, meaning that in a certain regime of intense field strengths the probability for ionizing an originally bound hydrogen atom by a pulsed oscillating field decreases rather than increasing, as one might expect, with increasing intensity of the external field (see also [23, 24]).

While there have been various attempts to understand which physical mechanism or mechanisms cause intense field stabilization, the studies to date indicate that the behaviour of ionization rates can largely be understood, at least qualitatively, in the framework of a classical description [19], even in the one-dimensional approximation [25]. Since ionization involves a transition into scattering states and since the classical dynamics of an atom in an oscillating field is largely chaotic, a basic understanding of the features of *chaotic scattering* is important for understanding the physics of the ionization process. Chaotic scattering is a topic of great current interest within the field of nonlinear dynamics; a focus issue of the journal "Chaos" contains a comprehensive account of recent developments in this subject [26].

In these circumstances there are at least two reasons for analyzing in some detail the predictions of the classical models describing an atom in an oscillating field. Such analyses may yield clues to the mechanisms underlying the ionization process in real atoms. Moreover the study of chaotic scattering in atomic systems may enrich the frequently rather abstract discussions in this field by showing up how features of chaotic scattering in realistic physical systems are similar to, and how they differ from, "generic" features observed in the usually studied abstract model systems.

The present article reviews recent work on chaotic scattering in simple (one-electron) atoms under the influence of an oscillating external field and is organised as follows. Section 2 briefly recalls the elements of chaotic scattering as observed in a "generic" model system. Section 3 describes recent results derived for a one-dimensional hydrogen atom subjected to an oscillating external field with a sinusoidal time dependence or to an external field consisting of a periodic sequence of instantaneous kicks. Particular attention is paid to the fractal structure of the set of unstable trapped orbits and to the question of whether ionization probabilities decay exponentially or as a power law in time. Section 4 shows how results obtained for a one-electron atom perturbed by a time-dependent external field may be related to features of chaotic scattering observed in unperturbed atoms with at least two electrons.

2 "Generic" Chaotic Scattering

In a standard scattering situation a physical system undergoes a transition from an asymptotic incoming state at large negative times t to an asymptotic outgoing state at large positive times due to a transient interaction. We are interested in how the outgoing state depends on the incoming state. In the classical scattering of a particle by a localized potential the essential information is contained in the deflection function describing the dependence of the angle of deflection on the impact parameter. For an electron subjected to an attractive Coulomb potential and an oscillating external field the asymptotic energy of Coulombic motion of the electron in the reference frame oscillating with the external field is one important observable, and a generalized deflection function can be defined which expresses the outgoing energy as a function of the incoming energy and the phase of the incoming motion relative to the oscillations of the external field [27, 28].

Chaotic scattering can occur if there is a set of (unstable) trajectories which are forever trapped by the interaction responsible for the scattering. Asymptotic incoming states which converge into such trapped trajectories for $t \to \infty$ are said to lie on the *stable manifold* of the trapped set whilst its *unstable manifold* consists of asymptotic outgoing states which approach the trapped set for $t \to -\infty$. The stable manifold of the trapped set obviously defines singularities in the deflection function – an incoming state deviating marginally from the stable manifold will approach a trapped trajectory very closely but eventually exit the interaction region and approach an asympotic state close to the unstable manifold. Small changes of the incoming state can drastically change the time the trajectory spends close to a trapped trajectory, and hence the outgoing state depends extremely sensitively on the precise parameters of the incoming state.

The structure of the scattering singularities can be investigated by studying the deflection function or the time delays suffered by scattered trajectories in comparison with "free trajectories" unaffected by the scattering interaction. Alternatively one can study the survival times of trajectories starting in the interaction region of phase space at time $t = 0$. Each such trajectory together with the time reversed trajectory starting at the same position with reversed momenta make up a scattering trajectory, and the time delay of the scattering trajectory is the sum of the two associated survival times. For an ensemble of trajectories initiated in the interaction region a *survival probability* $N(t)$ can be defined as the fraction of trajectories with survival times greater than or equal to t. For an atom perturbed by an external field, the survival time of a trajectory corresponds to the *ionization time*, i.e. the time it takes for an electron to leave the interaction region for good; in this case the survival probability of an ensemble of trajectories is just the fraction of non-ionized atoms.

The generic features of chaotic scattering are easily understood for the widely studied example of scattering of a point particle by three reflecting discs in a plane as illustrated in Fig. 1 [29–31]. The trapped trajectories are those which bounce around between the three discs and never find the gaps to escape. Unless the separation of the discs relative to their radius becomes too small, each infinite sequence of disc labels "1", "2" and "3" excluding immediate repetitions uniquely defines a trapped trajectory, so the set of trapped trajectories is uncountable. The stable manifold of the trapped set is defined by those impact parameters for which an incoming trajectory converges to a trapped trajectory for $t \to \infty$. The complement of the stable manifold of the trapped set consists of those trajectories which pass through the arrangement of the three discs without hitting any of them, those which find a gap to escape after being reflected once by a disc, those which are reflected twice before escaping, those which escape after three reflections, and so on. The dashed line in Fig. 1, for example, is a trajectory which escapes after six reflections. We obtain the set of impact parameters defining the stable manifold by successively excluding the finite intervals corresponding to escape after zero, one, two, three, etc. reflections. The number of finite intervals of impact parameters of trajectories undergoing at least $n + 1$ reflections is twice the number of intervals undergoing at least n reflections, because after the n-th reflection of a trajectory, e.g. by disc 2, there are two discs qualifying for the $(n + 1)$-st reflection, namely disc 1 and disc 3. The set of impact parameters leading to trapped trajectories and its

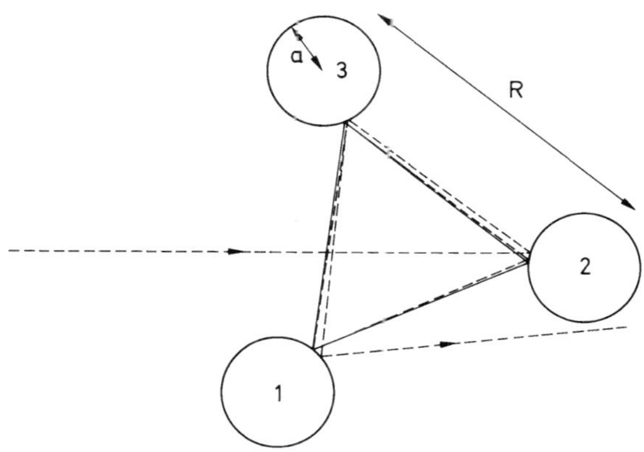

Fig. 1. Illustration of scattering by three discs of radius a, whose centres are separated by the distance R. The *solid line* is the periodic trapped trajectory labelled $\overline{123}$. The *dashed line* is a scattering trajectory which escapes from the interaction region after six reflections by a disc

construction are reminiscent of the standard *Cantor set*, which is constructed from the interval $[0, 1]$ by excluding the interval $]1/3, 2/3[$ in the first step and successively excluding the middle third from the resulting intervals in each subsequent step.

The standard Cantor set is a self-similar fractal with *fractal dimension* $d = \log 2/\log 3$. The fractal dimension d of a self-similar set is given by the dependence on ε of the number $v(\varepsilon)$ of finite intervals (or squares or cubes, etc.) of length ε needed to cover the set in the limit $\varepsilon \to 0$,

$$v(\varepsilon) \overset{\varepsilon \to 0}{\propto} \left(\frac{1}{\varepsilon}\right)^d. \tag{1}$$

In the construction of the standard Cantor set the total length of the intervals surviving after the n-th step is $(2/3)^n$ corresponding to a survival probability

$$N(n) = \exp(-\gamma n), \quad \gamma = -\log(2/3). \tag{2}$$

In particular, the measure of the Cantor set obtained in the limit $n \to \infty$ is zero, even though the set is uncountable. The rate coefficient γ defining the exponential decay of the survival probability in Eq. (2) is related to the fractal dimension d of the Cantor set by

$$\gamma = \left(\frac{1}{d} - 1\right) \log 2. \tag{3}$$

This relation also holds for Cantor sets defined by exclusion of other fractions than one third from the intervals in each step. The factor log 2 on the right of Eq. (3) has its origin in the fact that each step in the construction of the Cantor set involves a doubling of the number of surviving intervals.

Although this article concentrates on classical aspects of chaotic scattering, it should be mentioned that quantum mechanical manifestations of chaotic scattering have been studied by several authors (for a brief summary see [32]), in particular by Blümel [33] and Smilansky [34]. Emphasis is generally on the behaviour of the S-matrix, its eigenvalues and its energy autocorrelation function, which is semiclassically related to the Fourier transform of the scattering time delay function [35, 36]. Semiclassical methods based on periodic trapped orbits and creeping orbits, which follow the reflecting surfaces for some time, have been used to derive resonant poles of the S-matrix in the two-disc and the three-disc scattering problem by Wirzba [37] and Vattay et al. [38].

In the three-disc scattering problem, the set of impact parameters defining the stable manifold of the set of trapped trajectories is a fractal set of measure zero with a fractal dimension d depending on the separation of the discs relative to their radius. The scattering time delay of a trajectory is approximately proportional to the number of reflections it undergoes before escaping from the interaction region. The survival probability is the fraction of trajectories $N(n)$ undergoing at least n reflections, and it decays exponentially with n, $N(n) \propto \exp(-\gamma n)$, as illustrated in Fig. 2. The rate coefficient γ also depends on the ratio of the radius of the three discs to their separation and is in fact related

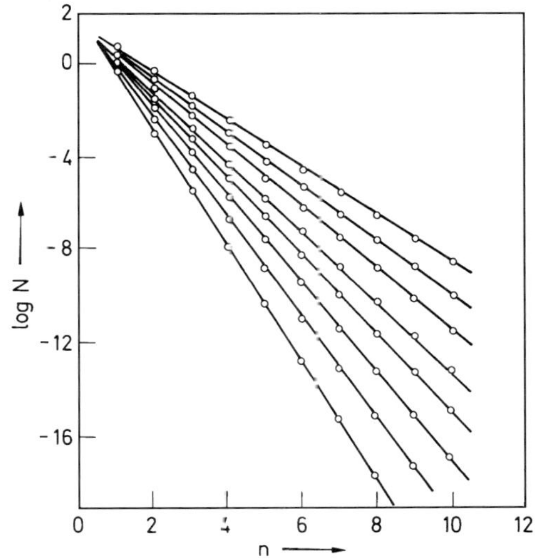

Fig. 2. Survival probabilities in the three-disc scattering system for various values of the ratio of the radius a of the discs to the separation R of their centres. The lines correspond from top to bottom to the following values of the parameter $2\sqrt{3}a/R$: 1.0, 0.9, 0.8, 0.7, 0.6, 0.5, 0.4, 0.3. (From [29])

to the fractal dimension d of the trapped set via Eq. (3) [29] just as for the standard Cantor set.

The three-disc system is an example of a *hyperbolic* scattering system. For a scattering system to be hyperbolic it must contain an uncountable set of trapped trajectories which are all unstable, i.e. the Liapunov exponent describing the rate of exponential growth of infinitesimal deviations from a given trajectory is positive for each trapped trajectory. Hyperbolicity also requires that the trapped trajectories are *uniformly unstable*, meaning that there is a finite positive lower bound to all Liapunov exponents [39]. These conditions are fulfilled for the three-disc system [29].

A chaotic scattering system is nonhyperbolic if it contains some stable periodic orbits surrounded by Kolmogorov-Arnold-Moser tori (KAM tori) upon which the dynamics evolve periodically or quasi-periodically. Such stable islands in the interaction region of phase space do not have stable and unstable manifolds extending into the asymptotic region, and hence they are not touched by the scattering trajectories. The presence of interspersed chains of islands of stability and remnants of KAM tori may however influence scattering trajectories passing close by and affect the fractal structure of the scattering singularities and the behaviour of the survival probability $N(t)$. Meiss and Ott [40] discussed a scenario in which the transport of scattering trajectories through the interaction region is slowed down by remnants of KAM tori, and found an algebraic (i.e. power-law) decay rather than an exponential decay of the survival probability in such a situation. Lau et al. [41] have conjectured that in chaotic scattering systems which are nonhyperbolic due to the presence of KAM tori, the fractal dimension of the set of scattering singularities is an integer, whilst it is non-integer in hyperbolic systems.

Such a connection between an integral dimension of the set of scattering singularities and a power-law decay of survival probabilities seems compatible with the simple relation at Eq. (3) valid for the standard Cantor set and in the three-disc scattering problem. Explicit investigations of more realistic atomic systems involving smooth long-ranged Coulomb potentials do however reveal that more complicated situations can occur. Examples are given in the following sections.

3 One-Electron Atom in an Oscillating External Field

The (non-relativistic) Hamiltonian for a hydrogen atom in an oscillating electric field is (in atomic units)

$$H(\boldsymbol{r}, \boldsymbol{p}; t) = \frac{\boldsymbol{p}^2}{2} - \frac{1}{r} + \varepsilon x f(\omega t), \tag{4}$$

where f is periodic with period $T = 2\pi/\omega$,

$$f(\omega t \pm 2\pi) = f(\omega t). \tag{5}$$

The last term in Eq. (4) assumes a field linearly polarized in the direction of the x-axis. To describe e.g. a circularly polarized sinusoidal field, this term should be replaced by $\varepsilon[x \sin(\omega t) \pm y \cos(\omega t)]$.

If we measure time in units of $t_0 = 1/\omega$, length in units of $a_0 = t_0^{2/3}$ and momentum in units of $p_0 = a_0/t_0 = t_0^{-1/3}$, then the Hamiltonian (4) may be rewritten in terms of the *scaled variables* $\tilde{\boldsymbol{r}} = \boldsymbol{r}/a_0, \tilde{\boldsymbol{p}} = \boldsymbol{p}/p_0$ and $\tilde{t} = t/t_0$ as

$$\omega^{-2/3} H \stackrel{\text{def}}{=} \tilde{H}(\tilde{\boldsymbol{r}}, \tilde{\boldsymbol{p}}; \tilde{t}) = \frac{\tilde{\boldsymbol{p}}^2}{2} - \frac{1}{\tilde{r}} + \tilde{\varepsilon}\tilde{x} f(\tilde{t}), \tag{6}$$

where $\tilde{\varepsilon}$ is the *scaled field strength*

$$\tilde{\varepsilon} = \varepsilon \omega^{-4/3}. \tag{7}$$

Thus the phase space structure of the classical dynamics depends not on field strength ε and frequency ω independently, but only on the scaled field strength defined by Eq. (7). Scaled energies $\tilde{E}$ and scaled actions $\tilde{I}$ derived from the equations of motion in the scaled variables are related to the energies E and actions I in atomic units by

$$\tilde{E} = \omega^{-2/3} E, \quad \tilde{I} = \omega^{1/3} I. \tag{8}$$

In particular, for negative energies the action integral $I_C = \oint \boldsymbol{p}\, d\boldsymbol{r}$ of unperturbed Coulombic oscillations corresponds to the principal quantum number n_0, which is related to the Coulomb energy via $E_C = -1/(2n_0^2)$. We may identify the corresponding scaled action $\tilde{I}_C = \oint \tilde{\boldsymbol{p}}\, d\boldsymbol{r}$ with a *scaled quantum number* $\tilde{n}_0$ related to n_0 by

$$\tilde{n}_0 = \omega^{1/3} n_0. \tag{9}$$

These scaling properties relate the classical dynamics in different regimes of field strengths and frequencies. For example, the classical dynamics evolving out of a hydrogenic bound state with an energy corresponding to quantum number n_0 in a field of frequency one atomic unit ($\omega = 1$ au $\Leftrightarrow$ $\omega/2\pi = 0.6579 \times 10^{16}$ Hz) and a field strength $\varepsilon = 1$ au ($= 5.142 \times 10^9$ V/cm) is, except for a scaling transformation, the same as the dynamics in a microwave field of 10 GHz corresponding to $\omega = 1.520 \times 10^{-6}$ au at a field strength $\varepsilon = \omega^{4/3} = 1.747 \times 10^{-8}$ au ($= 89.8$ V/cm), if we start from an initial bound state with quantum number $\omega^{-1/3} n_0 \approx 87 n_0$.

In a simple one-dimensional ansatz for a hydrogen atom in a sinusoidal electric field the Hamiltonian (4) reduces to

$$H(x, p; t) = \frac{p^2}{2} - \frac{1}{x} + \varepsilon x \sin(\omega t). \tag{10}$$

The classical dynamics generated by the Hamiltonian (10) and the relation to chaotic scattering have been studied in some detail by Wiesenfeld [42], who found an algebraic behaviour of survival probabilities in this system [27]. Special attention was given to the asymptotic states relevant for the scattering process [27, 28]. The incoming asymptote of a scattering trajectory is characterised by the asymptotic energy E of the Coulombic motion of the electron in the reference frame quivering with the amplitude $x_q = \varepsilon/\omega^2$ in the oscillating field, and by the phase difference ψ between the oscillating field and the Coulombic motion, the reference moment being taken at the point of closest approach of the electron. Figure 3 shows the deflection function $E_{\mathrm{out}}(E_{\mathrm{in}}, \psi_{\mathrm{in}})$ calculated for a fixed incoming energy $E_{\mathrm{in}} = 0.008$ au as function of ψ_{in} for a field strength parameter $\tilde{\varepsilon} \approx 0.1$. Note that the scaling behaviour of the scattering singularities is not strictly self-similar. The relative lengths of the intervals where the deflection function is smooth decrease dramatically when going from middle panel to the lowest panel. Wiesenfeld attributed this behaviour to islands of stability consisting of KAM tori surrounding tightly bound stable periodic orbits or to weakly unstable Rydberg states.

The role of KAM tori in the (classical) microwave ionization of hydrogen atoms has also been studied by Lai et al. [43]. These calculations are again based on the one-dimensional Hamiltonian (10), but the x-coordinate is restricted to the semi-axis of positive values. Figure 4 shows phase space portraits for field strengths $\tilde{\varepsilon} = 0.015$ (a) and $\tilde{\varepsilon} = 0.019$ (b). They are obtained by plotting the values of the scaled action variable $\tilde{I}$ and the conjugate angle variable Θ of trajectories at discrete times separated by the period $T = 2\pi/\omega$ of the external field ("stroboscopic map"). The action variable $\tilde{I}$ is related to the (scaled) energy $\tilde{E}$ of the electron in the Coulomb field by $\tilde{E} = -1/(2\tilde{I}^2)$, cf. Eq. (8). The dynamics are clearly stable for low values of $\tilde{I}$ and mainly

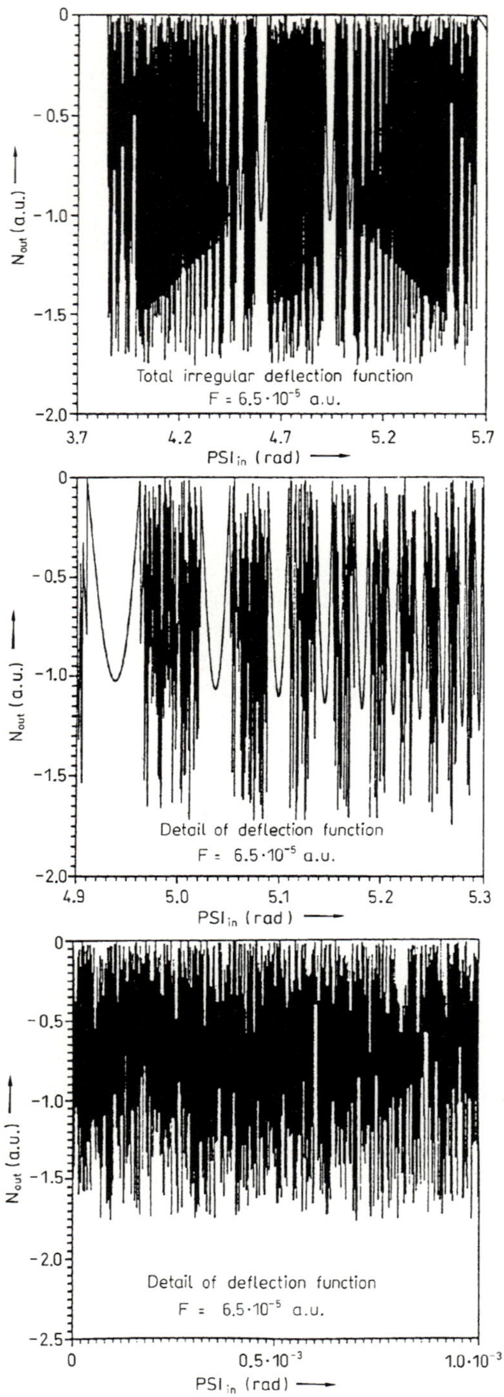

Fig. 3. Deflection function for an incoming electron with an asymptotic energy of Coulombic motion $E_{\text{in}} = 0.008$ au for a frequency $\omega = 0.004$ au and a scaled field strength $\tilde{\varepsilon} = 0.1$. The ordinate label is related to the outgoing energy E_{out} via $N_{\text{out}} = -E_{\text{out}}/0.008$ in this case. ψ is the phase of the Coulombic motion relative to the oscillating external field at the point of closest approach. In the bottom panel the phase ψ is given relative to $\psi_0 = 4.12$. (From [28])

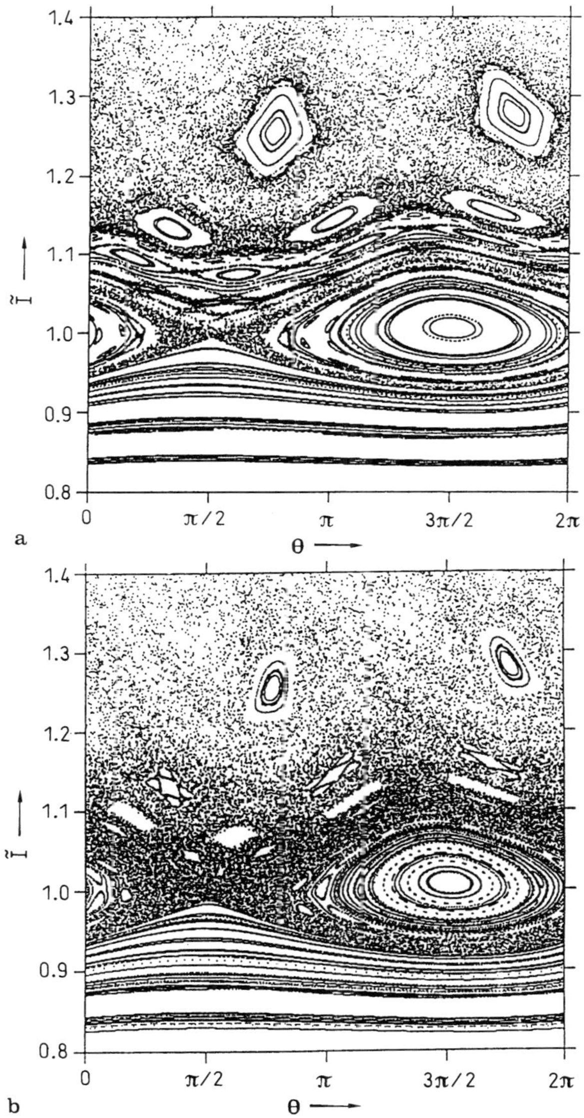

Fig. 4a, b. Phase space portraits for trajectories generated by the Hamiltonian (10) (on the semi-axis of positive x-values): **a** for $\tilde{\varepsilon} = 0.015$; **b** for $\tilde{\varepsilon} = 0.019$. The ordinate label $\tilde{I}$ is the scaled action which is related to the scaled energy $\tilde{E}$ of bound Coulombic motion of the electron by $\tilde{E} = -1/(2\tilde{I}^2)$, and Θ is the conjugate angle variable. (From [43])

unstable for larger values of $\tilde{I}$. Interspersed layers of stable periodic orbits surrounded by KAM tori appear in the intermediate region around $\tilde{I} \approx 1$. Increasing the field strength leads to continued breaking-up of KAM tori and the expansion of the chaotic region of phase space.

The representation in Fig. 4 requires negative energies of the electron in the Coulomb potential. As is common practice in such investigations, Lai et al. don't explicitly treat the asymptotic states of the electron, but study the behaviour of its binding energy or its (scaled) action $\tilde{I}$ as a function of (scaled) time. When $\tilde{I}(\tilde{t})$ increases beyond a certain large limit $\tilde{I}_c$ the atom is considered to ionize, and the time this takes is the survival time of the trajectory. Trajectories initiating from the stable region or from a stable island are periodic or quasiperiodic and hence all survive indefinitely. In the chaotic region there is a set of trapped unstable trajectories as is typical for a chaotic scattering system, and trajectories initiated here can have finite or infinite ionization times. Figure 5 shows doubly logarithmic plots of the survival probabilities for trajectories initiated with $\tilde{I} = 1.4$ and uniformly distributed initial angles Θ at the two field strengths of Fig. 4. The fall-off is clearly algebraic rather than exponential for long times. Lai et al. attribute the algebraic decay to sojourns of the trajectories in the vicinity of KAM tori surrounding stable periodic orbits, as in the scenario of Meiss and Ott [40]. At the slightly higher field in Fig. 5b, there is a crossover to a weaker power in the decay near a crossover time $\tilde{t} = 3$. Lai et al. attribute this to some of the trajectories leaking through broken tori into phase space regions hardly accessible at the lower field strength. If this interpretation of Lai et al. is correct, we may expect further crossovers at still larger times as very long-lived trajectories probe finer and finer leaks in broken KAM tori.

The studies of Wiesenfeld [28] and Lai et al. [43] on the classical dynamics of a one-electron atom in a sinusoidal external field provide a physically realistic example in which the presence of KAM tori surrounding stable periodic orbits leads to deviations from the generic behaviour characteristic of a hyperbolic scattering system as discussed in Sect. 2. Although this system (10) seems simple, further studies illuminating the mathematical structures behind the scattering process, e.g. calculation of the Liapunov exponents of the unstable trapped orbits and the fractal dimension of the trapped set, have yet to be performed.

A more detailed investigation has been carried out [44, 45] for the periodically kicked hydrogen atom, in which the sinusoidal dependence of the external field in Eq. (10) is replaced by a periodic sequence of instantaneous kicks,

$$H(x, p; t) = \frac{p^2}{2} - \frac{1}{x} - \varepsilon x \sum_{k=0}^{\infty} \delta(\omega t - 2k\pi), \quad x > 0, \tag{11}$$

with period $T = 2\pi/\omega$. For positive values of the kick strength ε, the change of momentum due to each kick is positive, i.e. the external field always kicks the electron away from the centre of the Coulomb force at $x = 0$. For negative ε the kicks are always directed towards $x = 0$. Alternating kick strengths, $\varepsilon_k \propto (-1)^k$, may seem the most reasonable simulation of the more realistic sinusoidal time dependence. The case of constant positive kick strength ε is, however, of particular theoretical interest, because the phase space in this case contains no stable periodic orbits and no KAM tori. It can in fact be shown that all periodic orbits are uniformly unstable for constant positive kick strength [45].

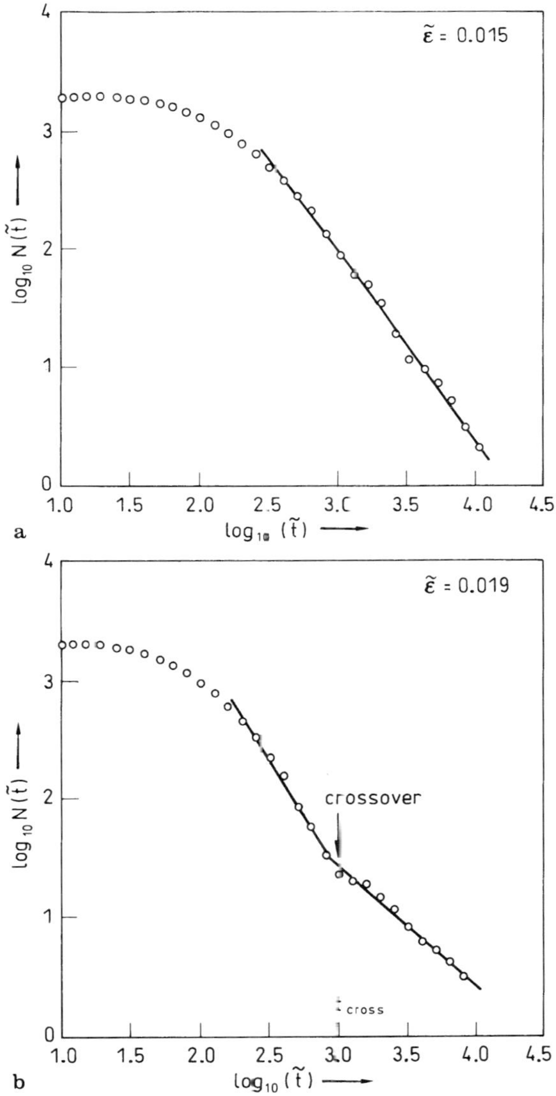

Fig. 5a, b. Survival probability $N(\tilde{t})$ of trajectories initiated in the chaotic regions of phase space as shown in Fig. 4, namely with $\tilde{I} = 1.4$ and uniformly distributed angles Θ: **a** for $\tilde{\varepsilon} = 0.015$; **b** for $\tilde{\varepsilon} = 0.019$. (From [43])

The dynamics of systems in which the non-integrable part of the Hamiltonian acts only in instantaneous kicks is particularly easy to visualize, because in between kicks the evolution is integrable and can be labelled by the appropriate integrals of motion, such as the energy of the electron in the Coulomb field in the case of the kicked hydrogen atom. For positive kick strength the survival time of a trajectory starting from the bound regime, i.e. with a negative energy, can be unambiguously defined as the number of periods or kicks it undergoes before it acquires a positive energy. The electron's momentum is always positive after the kick leading to positive energy [44], and hence this kick unambiguously defines the instant of ionization. The survival times impose a fractal structure on phase space as illustrated in Fig. 6 for scaled field strength $\bar{\varepsilon} = 1$. Here n stands for the scaled action labelled $\tilde{I}$ in Fig. 4, and Θ is the conjugate angle variable. The white region corresponds to initial conditions of trajectories which ionize after the first kick. The red, green, blue, yellow, violet, and turquoise stripes mark the starting points of trajectories surviving one to six further periods respectively before ionization. The middle and lower panels are magnifications of the regions marked by a black square in the preceding panel.

The fractal structure in phase space is obvious in Fig. 6. The fractal properties of the set of trapped trajectories can be studied by looking at a one-dimensional cut through the stripes in phase space. The initial conditions of the trapped set are those points which survive indefinitely, i.e. on no level of magnification belong to one of the uniformly coloured stripes, in analogy to the standard Cantor set which contains those points of the interval [0, 1], which in no step of construction belong to one of the excluded middle thirds. It turns out that the trapped set for kicked hydrogen has a *broken scaling symmetry*, it is not strictly self-similar, its measure is zero, but its fractal dimension is unity. This may be compared with a Cantor set constructed by excluding not a constant fraction such as one third in each step, but by excluding a varying fraction such as $1/(\nu + 2)$ in the ν-th step. The Cantor set obtained in this way also has measure zero and its fractal dimension is unity.

The broken scaling symmetry of the trapped set for the periodically kicked hydrogen atom is accompanied by a power-law decay of the survival probabilities as in the case of a sinusoidal field studied by Wiesenfeld and Lai et al. This is illustrated in Fig. 7 showing a doubly logarithmic plot of the survival probabilities for initial conditions from five different cuts through phase space in Fig. 6 corresponding to different fixed values of the scaled action $\tilde{I} \equiv n_0$ and uniformly distributed angles θ. In the case of the sinusoidal field, broken scaling symmetry (Fig. 3) and algebraic decay of survival probabilities (Fig. 5) were attributed to the presence of interspersed islands of stability in phase space. For the present example of kicked hydrogen, where all periodic orbits are uniformly unstable, the origin of these deviations from the generic behaviour of hyperbolic systems as described in Sect. 2 is different, and can be understood in the following way: If the electron is moving towards the origin at the instant of the kick, it is generally decelerated by the kick, while an electron moving away is always accelerated and may ionize. For this reason, the trajectories not ionizing after a given kick contain a bias towards lower actions (i.e. tighter binding) compared with the ensemble of trajectories before the kick. Since the probability for ionization at a subsequent kick decreases with decreasing action of the trajectory before the kick, a progressively decreasing fraction of trajectories is ionized at each step.

The dynamics of evolution become somewhat more complicated if the external field is not strictly periodic but rises and falls during a finite switch-on and switch-off time respectively, as will realistically be the case for an atom

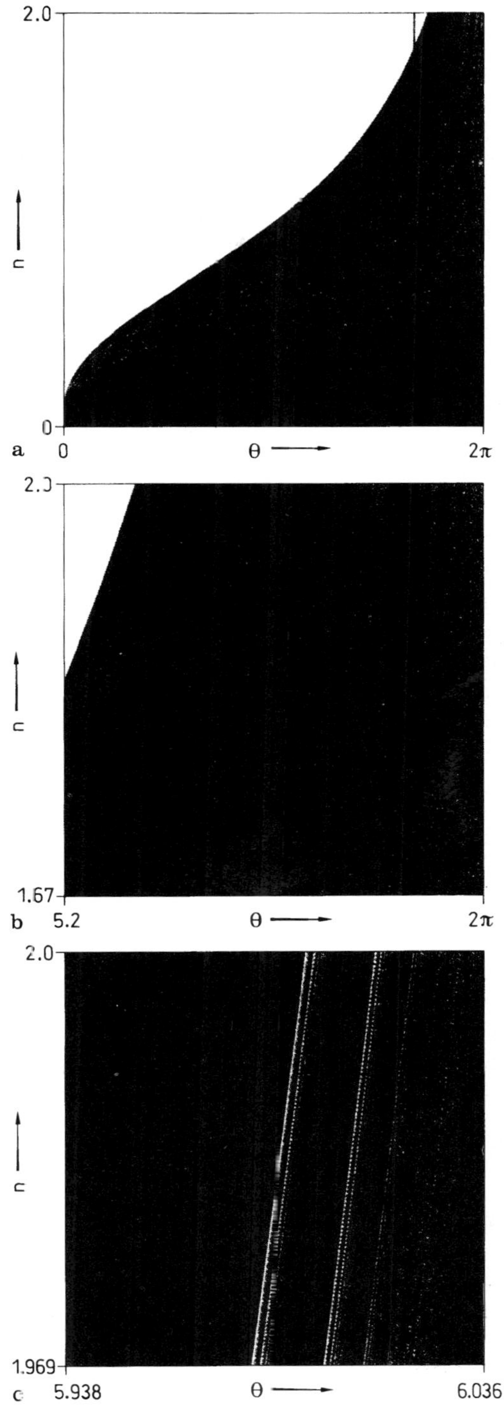

Fig. 6a–c. Survival times of trajectories for the one-dimensional hydrogen atom subjected to periodic kicks (Eq. (11)) with constant positive scaled kick strength $\tilde{\varepsilon} = 1$. The white region shows initial conditions of trajectories which ionize after the first kick. The red, green, blue, yellow, violet and turquoise stripes mark the starting points of trajectories surviving one to six further periods. The middle and lower panels are magnifications of the regions marked by a black square in the preceding panel. (From [44])

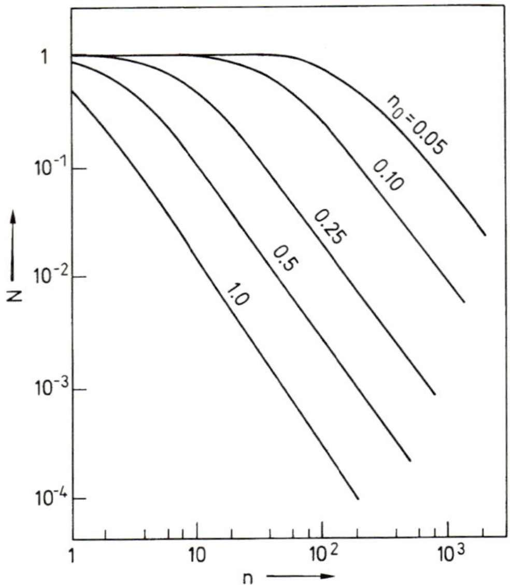

Fig. 7. Doubly logarithmic plot of the survival probabilities $N(n)$ in kicked hydrogen (Eq. (11)) as functions of the number n of kicks for scaled field strength $\tilde{\varepsilon} = 1$. The various curves are obtained with initial conditions on different cuts through phase space corresponding to fixed values of the scaled action $\tilde{I} = n_0$ and uniformly distributed angles. (From [45])

exposed to a pulse of laser or microwave radiation. Gebarowski and Zakrzewski [46] have observed algebraic decay of survival probabilities in classical calculations for a hydrogen atom subjected to a finite and sufficiently long pulse, and find that the power defining the decay rate depends on the shape of the pulse and also, for a two-dimensional atom and a circularly polarized field, on the angular momentum of the initial bound state.

Beeker and Eckelt [62] studied an interesting modification of the usual kicked atom situation by introducing a kick strength which is modulated not in time but in space,

$$H(x, p; t) = \frac{p^2}{2} - \frac{\varepsilon}{(1 + x^2)^2} \sum_{k=-\infty}^{\infty} \delta(\omega t - 2k\pi). \tag{12}$$

In this ansatz there is no Coulomb potential but an attractive well of depth ε, which is switched on for infinitely short intervals after each period $2\pi/\omega$. Since the potential dies out for $x \to \pm\infty$, the asymptotic states are easily identified: they are defined by the asymptotic kinetic energy E of the particle and its phase $2\pi\tau$ relative to the periodic oscillations of the potential in time.

In Eq. (12) the kicks are always directed towards the origin $x = 0$. A typical scattering trajectory starts at $x = -\infty$ with a positive asymptotic energy and passes the origin $x = 0$ at least once, but perhaps several times, and its scattering time delay can be unambiguously defined as the time between the first and the last zero of the trajectory. The dynamics generated by the Hamiltonian (12) feature stable periodic orbits and a rich KAM scenario for sufficiently weak field strengths ε and a complete absence of KAM tori for sufficiently large ε.

There is a Cantor set of trapped trajectories which show up in the deflection functions or in a phase space portrait of the scattering time delays or survival times. This is indicated in Fig. 8 showing the phase space structure at a field strength $\varepsilon = 2$, where there are no islands of stability. The initial conditions of those trajectories with exactly two zeros are marked black, the white regions in between correspond to trajectories with three or more zeros. A break-down of these regions according to the number of zeros reveals self-similar fractal structure [62].

The chaotic properties of scattering in this example can be studied e.g. by looking at the time delays Δt of the scattering trajectories $x(t)$ or at their numbers of zeros. The fraction N_n of trajectories with n zeros does actually fall off exponentially as a function of n, as expected for a self-similar structure of scattering singularities, see Fig. 9a. If, however, the fraction $N(\Delta t)$ of trajectories is plotted as a function of the time delay Δt, the fall-off is not exponential but algebraic, as illustrated by the doubly logarithmic plot in Fig. 9b. The reason for this lies in the occurrence of trajectories which almost escape after e.g. the n-th zero, but are eventually recaptured, so that there is a long time elapsing between the n-th and the $(n + 1)$-th zero. This simple example shows that algebraic decay of survival probabilities can be compatible with unbroken scaling of the trapped set when the "generation parameter" counting the generations of self similarity of the trapped set (number of zeros in this example) is not simply proportional to the time variable as in the example, Eq. (11), studied by Hillermeier et al. [45].

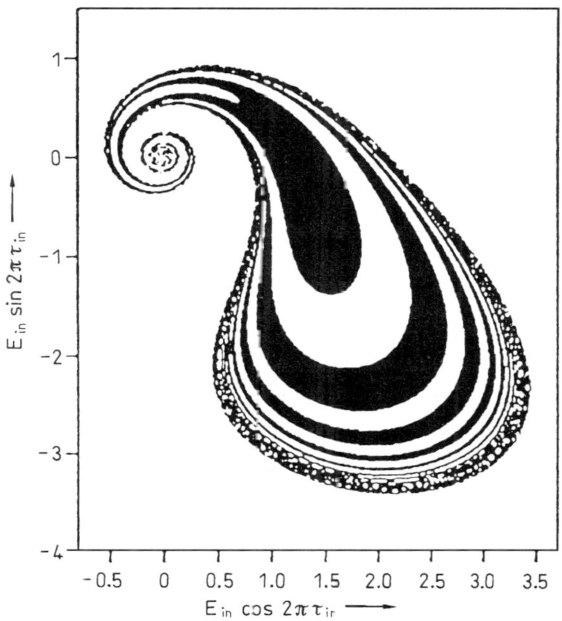

Fig. 8. Phase space for the dynamics generated by the Hamiltonian (12) for $\varepsilon = 2$. The initial conditions of scattering trajectories are defined by the asymptotic incoming energy E_{in} and a phase $2\pi\tau_{in}$ relative to the oscillating potential. Initial conditions leading to trajectories $x(t)$ with a total of two zeros are marked black. The white regions in between correspond to trajectories with three or more zeros (From [62])

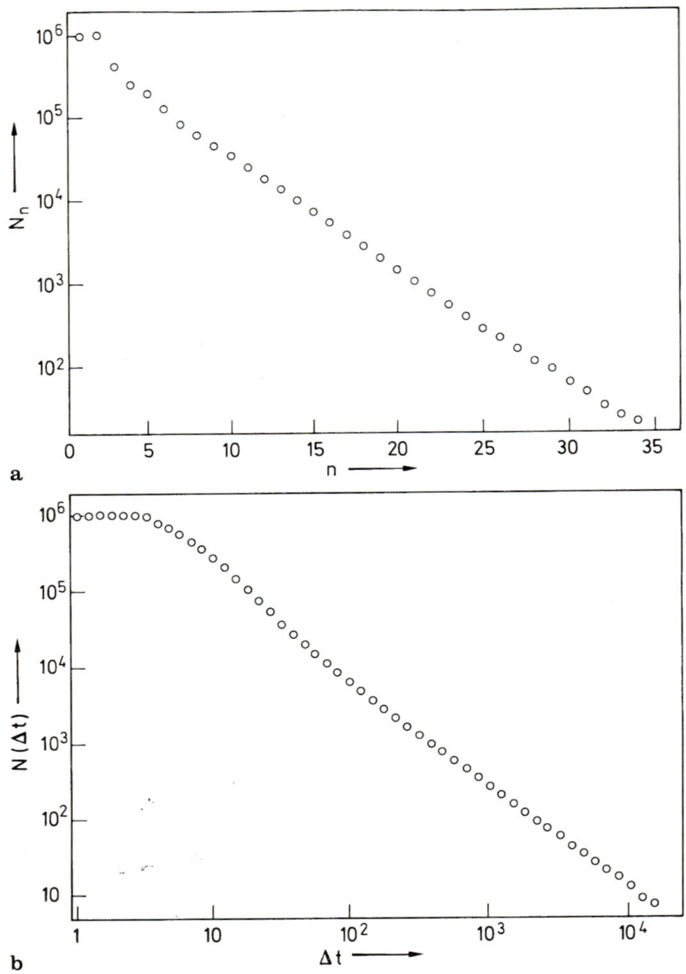

Fig. 9a, b. Survival probabilities N for an ensemble of initial conditions corresponding a uniform distribution of incoming energies E_{in} along a cut defined by $\sin 2\pi\tau_{\text{in}} = 0$, $\cos 2\pi\tau_{\text{in}} = 1$ in Fig. 8: **a** logarithmic plot of N as function of the number n of zeros of the trajectories revealing an exponential decay; **b** doubly logarithmic plot of N against the scattering time delay Δt revealing power-law decay

4 Chaotic Scattering in Two-Electron Atoms

In the examples discussed above, the integrability of the classical dynamics of a one-electron atom is destroyed by an external field. Integrability is also destroyed if the system consisting of one electron and an atomic nucleus is complemented by at least one further electron, even without any external field.

The (non-relativistic) Hamiltonian e.g. for an unperturbed two-electron atom (or ion) with a nucleus of infinite mass and charge number Z is (in atomic units)

$$H(\boldsymbol{r}_1, \boldsymbol{r}_2; \boldsymbol{p}_1, \boldsymbol{p}_2; t) = \frac{\boldsymbol{p}_1^2}{2} + \frac{\boldsymbol{p}_2^2}{2} - \frac{Z}{r_1} - \frac{Z}{r_2} + \frac{1}{|\boldsymbol{r}_1 - \boldsymbol{r}_2|}. \tag{13}$$

Since the potential in Eq. (13) is homogeneous, the classical dynamics generated by the Hamiltonian does not depend on the magnitude of the (conserved) energy E, except for a scaling transformation. As far as classical mechanics is concerned, there are only three essentially different energies, $E < 0$, $E = 0$ and $E > 0$. $E = 0$ defines the threshold for total break-up, where all electrons can escape to infinity. The scattering states obviously play a central role at positive and vanishing total energy, but they are also essential for understanding the classical dynamics at negative energies. In contrast to the quantum mechanical case, where there is a lower bound to the energy of an electron in a Coulomb potential, a classical electron can come arbitrarily close to the atomic nucleus and hence acquire an arbitrarily large negative energy. In doing so it may transfer the energy it loses while approaching the nucleus to another (initially bound) electron which can be excited into a scattering state, thus ionizing the atom. Ionization is always energetically possible, there is no ionization threshold. Classical atoms with at least two electrons are examples of non-integrable systems with scattering trajectories at any energy, and hence understanding the features of chaotic scattering is important for understanding their dynamics.

For total angular momentum zero the two-electron system (Eq. (13)) is defined by three independent coordinates, e.g. the distances r_1, r_2 of the electrons from the nucleus and the angle between $\boldsymbol{r}_1$ and $\boldsymbol{r}_2$. For non-vanishing total angular momentum there is a fourth coordinate corresponding to the angle between the angular momentum vector and the plane spanned by $\boldsymbol{r}_1$ and $\boldsymbol{r}_2$. Correspondingly, phase space has five or seven dimensions after energy conservation is accounted for. Investigations of the classical dynamics in atomic systems with such large dimensions are still rare. First studies of patterns of chaotic scattering in the classical helium atom were performed by Gu and Yuan [47] and by Yamamoto and Kaneko [48]. Considerable progress has been made in the study of the classical dynamics of the helium atom in one-dimensional approximations with only two spatial coordinates r_1, r_2 describing the distances of the electrons from the nucleus,

$$H(r_1, r_2; p_1, p_2; t) = \frac{p_1^2}{2} + \frac{p_2^2}{2} - \frac{Z}{r_1} - \frac{Z}{r_2} + V(r_1, r_2). \tag{14}$$

A configuration in which both electrons are confined to lie on a straight line through the nucleus and on opposite sides of the nucleus is described by an electron-electron interaction potential

$$V(r_1, r_2) = \frac{1}{r_1 + r_2} \tag{15}$$

in the Hamiltonian (14). A complementary model, in which both electrons are assumed to have spherical shape and which hence contains no angular correlations at all, is described by the potential

$$V(r_1, r_2) = \frac{1}{r_>}, \qquad (16)$$

where $r_>$ is the larger of the two radial coordinates r_1 and r_2.

The collinear model (Eq. (15)) has been successfully used in the semiclassical description of many bound and resonant states in the quantum mechanical spectrum of real helium [49–52] and plays an important role for the study of states of real helium in which both electrons are close to the continuum threshold [53, 54]. The quantum mechanical version of the spherical or *s*-wave model (Eq. (16)) describes the 1*sns* bound states of real helium quite well [55]. The energy dependence of experimental total cross sections for electron impact ionization is reproduced qualitatively in the classical version of the *s*-wave model [56] and surprisingly well quantitatively in a quantum mechanical calculation [57]. The *s*-wave model is less realistic close to the break-up threshold $E = 0$, where motion along the Wannier ridge, $r_1 = r_2$, is important.

The classical dynamics of collinear helium (Eq. (15)) and of *s*-wave helium (Eq. (16)) [56, 58] are very similar for negative energies. Asymptotic states corresponding to one bound electron and one incoming electron are labelled by the (positive) asymptotic energy of the incoming electron and the phase of its motion relative to the Coulomb oscillations of the bound electron. There is a Cantor set of trapped trajectories which are all unstable, there are no stable periodic orbits and no KAM tori. Quantitative studies of the phase space structure are somewhat easier in the *s*-wave model (16), because the equations of motion are separable in the regions $r_1 > r_2$ and $r_1 < r_2$, and the energies E_1, E_2 of the individual electrons are conserved within these regions. The two electrons only exchange energy when they meet on the diagonal $r_1 = r_2$ in coordinate space. The scattering time delay of a trajectory is unambiguously defined as the time between the first and the last encounter of the two electrons at $r_1 = r_2$, and the survival time or ionization time of a trajectory with initial conditions corresponding to both electrons bound ($E_1 < 0, E_2 < 0$) is the time which elapses until one electron acquires positive energy [58]. Figure 10 shows the ionization times of trajectories with initial conditions in the $r_1 - p_1$ section through the phase space at $r_2 = 0$. Yellow corresponds to short ionization times and blue to long ionization times as shown in the legend at the bottom of the figure.

The self-similar fractal structure is obvious in Fig. 10. On each scale there is one dominant yellow stripe, on the left in the uppermost panel and in the middle in the other two panels. This stripe corresponds to ionization after a certain number of oscillations of both electrons in the Coulomb field. The narrower yellow stripes accumulating towards the blue boundaries near the left- and right-hand edges of the panels are populated by trajectories in which one electron travels far from the nucleus with a very small negative energy, and hence returns, while the inner electron performs short Coulombic oscillations; each

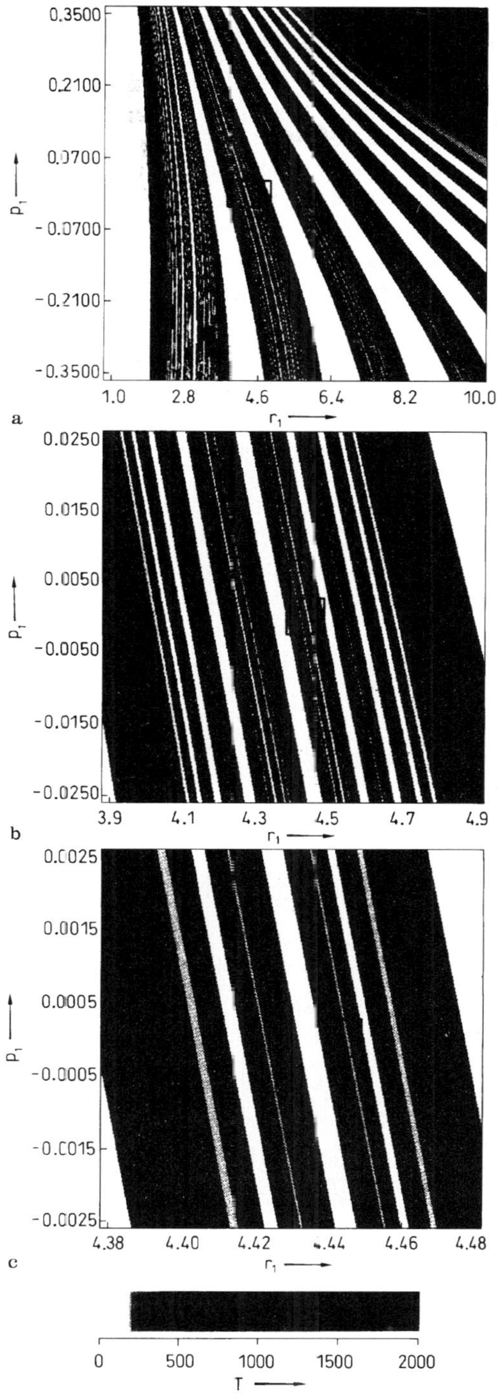

Fig. 10a–c. Ionization times in *s*-wave helium ($Z = 2$). The panels show the $r_1 - p_1$ section through the phase space at $r_2 = 0$. Initial conditions in the coloured region correspond to two bound electrons (i.e. both electrons have negative energy), and the colour is related to the ionization time T according to the legend at the bottom of the figure. The middle and lower panels are magnifications of the regions marked by a green rectangle in the preceding panel. (From [58])

successive yellow stripe corresponds to one further oscillation of the inner electron during such an excursion of the outer electron. On the other hand, moving from one generation of yellow stripes to the next, i.e. going from the central yellow stripe in the middle panel to the central yellow stripe in the lowest panel, corresponds to one or two additional encounters of the two electrons at $r_1 = r_2$ before ionization. (For a detailed account of the structures in Fig. 10 see [56, 58].)

The generation parameter defining the generation of ionizing trajectories in the self-similar structure in Fig. 10 is related to the number w of encounters of the two electrons at $r_1 = r_2$ rather than to the ionization time. This interpretation is confirmed in Fig. 11 which shows the density n of trajectories starting with initial conditions uniformly distributed in the middle panel of Fig. 10 as function of the number w of encounters of the two electrons and of the ionization time T. The density n is proportional to minus the derivative of the survival probability with respect to the relevant variable (w or T). The logarithmic plot in Fig. 11a reveals an exponential decay of the density, $n(w) \propto \exp(-0.27w)$, and hence also of the survival probability, as a function of the number of encounters of the two electrons, just as expected for a self-similar fractal set of trapped trajectories. The doubly logarithmic plot of the density of trajectories in Fig. 11b reveals a power-law decay of the density, $n(T) \propto T^{-1.82}$, and hence also of the survival probability,

$$N(t) = \int_t^\infty n(T)dT \propto t^{-0.82}, \tag{17}$$

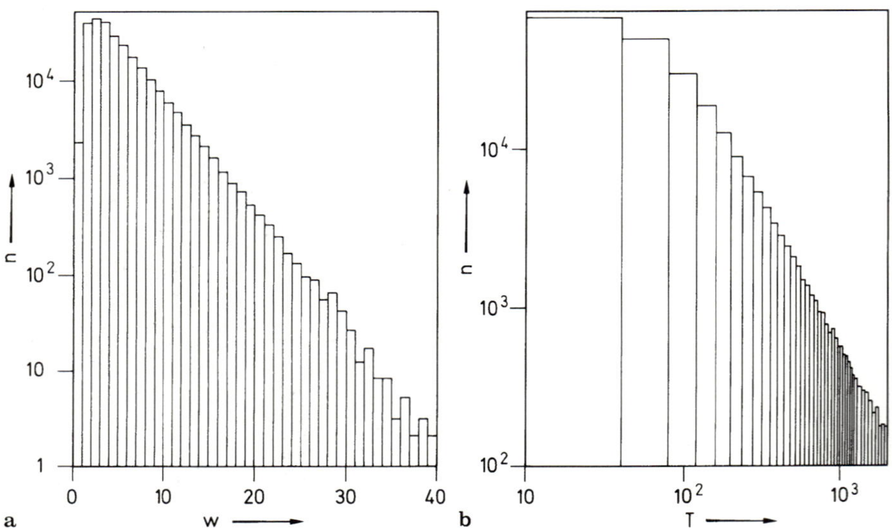

Fig. 11a, b. Density n of trajectories with initial conditions uniformly distributed in the part of phase space shown in the middle panel of Fig. 10: **a** as function of the number w of encounters of the two electrons; **b** as function of the ionization time T. Note that the plot is logarithmic in (a) and doubly logarithmic in (b). (From [58])

as a function of the ionization time for long times. The slower decay of the survival probability as function of time is due to the occurrence of long-lived trajectories in which the outer electron has a small negative energy and performs a slow Coulombic oscillation through a distant turning point in between two encounters with the other electron at $r_1 = r_2$. This is analogous to the situation in the model of Beeker and Eckelt (Eq. (12)) where a long time can elapse between two zeros for trajectories which only almost escape. Note that it is not the short or long ranged nature of the potential which is important here, but the possibility of arbitrary long times elapsing between two encounters or zeros.

The fractal dimension of the trapped set of trajectories of *s*-wave helium initiated on a horizontal cut through any one of the panels in Fig. 10 has been calculated by Handke [59] to be $d = 0.93$, which is clearly less than unity. This is consistent with the expectation for a strictly self-similar set as described in Sect. 2. It does not obey the simple relation (3) connecting the fractal dimension with the rate coefficient of the exponential decay of survival probabilities, which is not surprising, because Eq. (3) was derived for fractals in which the number of connected intervals surviving the construction process doubles in each step, whereas for *s*-wave helium illustrated in Fig. 10 each generation contains infinitely many more intervals (homogeneously coloured stripes) than the preceding generation.

It is interesting to compare the quantitative results for *s*-wave helium with the results obtained by Hillermeier et al. [45] for periodically kicked hydrogen (with constant positive kick strength) as discussed in Sect. 3. In both cases the trapped orbits are all unstable, there are no islands of stability in phase space. In kicked hydrogen the trapped orbits are uniformly unstable, but the survival probabilities decay algebraically and the fractal dimension of the trapped set is an integer. In *s*-wave helium the fractal properties of the trapped set seem somewhat closer to the expectations for a hyperbolic system as discussed in Sect. 2, the fractal dimension is non-integer and the survival probability decays exponentially as function of the number of encounters of the two electrons, although the decay as function of time is algebraic.

However, *s*-wave helium does not strictly fulfill the requirements of a hyperbolic scattering system, because the trapped orbits are not uniformly unstable. To see this consider the sequence of trapped periodic orbits in which one period contains one encounter on the line $r_1 = r_2$ in coordinate space, whereafter the outer electron has a small negative energy and performs a slow Coulombic oscillation through a distant turning point while the inner electron performs $n = 1, 2, 3, \ldots$ unperturbed oscillations. In the usual coding scheme [58] such orbits are labelled $\overline{- + + \cdots +}$ with one 'minus' sign and n 'plus' signs. All orbits in this sequence feel the interaction responsible for the instability at only one instant during each period and evolve integrably for other times. This is a particularly drastic realisation of the concept of *intermittent regions in phase space*, where the dynamics evolve integrably or almost integrably, as discussed by Dahlqvist [60]. The Liapunov exponent λ of a periodic orbit of period T is given [3] by $\lambda = (\log|\Lambda|)/T$, where $|\Lambda|$ is the instability factor describing the

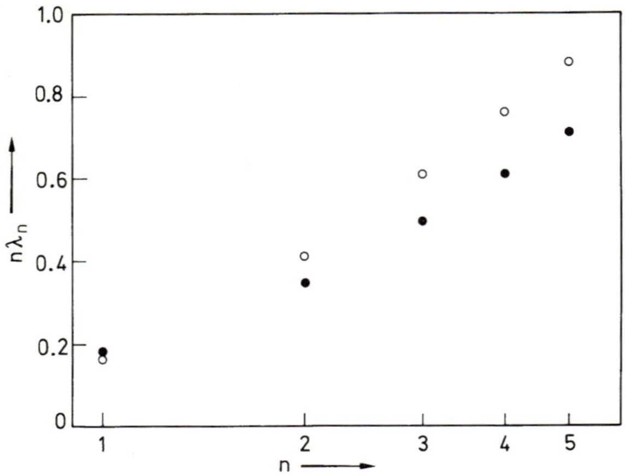

Fig. 12. Liapunov exponents λ_n of the unstable periodic orbits labelled $\overline{-++\cdots+}$, where each period contains one slow Coulombic oscillation of the farther electron and n Coulombic oscillations of the closer electron in between two crossings of the line $r_1 = r_2$ in coordinate space. The ordinate measures the products $n\lambda_n$ on a linear scale and the abscissa measures n on a logarithmic scale. The *solid dots* show the results for *s*-wave helium (Eq. (16)) [55], the *open circles* show the corresponding results for collinear helium (Eq. (15)) [52]. The linear behaviour for large n illustrates the proportionality of λ_n to (log n)/n. (Results are for charge $Z = 2$ in the Hamiltonian, Eq. (14))

[The stability exponents labeled u in Table 1 of Wintgen et al. [52] are actually the Liapunov exponents of the periodic orbits multiplied by their respective periods, $u = \lambda T$.]

maximal growth of infinitesimal deviations from the periodic orbit after one period. For the orbits $\overline{-++\cdots+}$ with a large number n of oscillations of the inner electron during each period, this factor behaves as a power of n, because the motion is integrable during these oscillations. The period T depends linearly on n, so the Liapunov exponents λ_n are proportional to log n/n for large n, as illustrated in Fig. 12, where the products $n\lambda_n$ are plotted against n on a logarithmic scale for the five orbits labelled $\overline{-+}$ to $\overline{-+++++}$ (filled dots). The dependence of $n\lambda_n$ on log n tends to become linear for large n. The same behaviour is observed for the corresponding orbits in collinear helium [52], where the motion of the two electrons is not exactly but almost integrable for $r_1 \gg r_2$ or $r_2 \gg r_1$ (open circles in Fig. 12).

5 Summary

Chaotic scattering is a topic of great current interest within the field of nonlinear dynamics. The majority of investigations on this topic have been performed for rather abstract model systems, usually featuring free particle motion reflected at sharply defined boundaries of various geometric shapes. Properties observed in such systems are regarded "generic", e.g. the self-similar fractal structure of the set of unstable trapped trajectories, the exponential decay of survival probabilities in hyperbolic scattering systems, the connection between hyperbolicity of the dynamics on the trapped set and a non-integer value of its fractal dimension.

One-electron atoms subjected to a time-dependent external field provide physically realistic examples of scattering systems with chaotic classical dynamics, and the study of such systems can substantially enrich the discussion of chaotic scattering in general. The examples discussed in Sect. 3 reveal several features which are not obviously compatible with what is usually regarded as generic and deserve further investigation. The broken scaling symmetry of the trapped set and the algebraic decay of survival probabilities for a hydrogen atom in a sinusoidal external field (Eq. (10)) seems accounted for by the occurrence of interspersed KAM islands of stability, whereby the scattering system is nonhyperbolic. The fact that the power describing the algebraic decay of the survival probability can change at large times (Fig. 3b) could force us to modify our concept of algebraic decay for large times. Broken scaling symmetry and algebraic decay are also observed for the periodically kicked hydrogen atom (Eq. (11)), but they are more difficult to explain, at least for the case of constant positive kick strength, because there are no KAM islands and all periodic trapped orbits are uniformly unstable in this case.

The simple model (12) studied by Beeker and Eckelt [62] demonstrates that unbroken scaling of the fractal set of the trapped unstable trajectories can be compatible with algebraic decay of the survival probability, if the generation parameter counting successive generations in the construction of the trapped set is not simply proportional to the ionization time. The survival probability actually decays exponentially as a function of the generation parameter in this example, and the slower algebraic decay in time can be understood as due to trajectories along which arbitrarily long times can elapse between two successive values of the generation parameter.

Another class of physically realistic examples of chaotic scattering systems consists of atoms with at least two electrons in the absence of external fields. Quantitative results are available for two-electron atoms described in one-dimensional models, such as the s-wave model (Eqs. (14) and (16)) or the collinear model (Eqs. (14) and (15)). All trapped orbits are unstable, but not uniformly unstable in these systems, so they are not strictly hyperbolic. Nevertheless, unbroken scaling symmetry and a non-integer fractal dimension of the trapped set have been reported for s-wave helium, and this is accompanied by exponential decay of the survival probability as function of the number of encounters of the two electrons. The survival probability decays algebraically as a function of time due to the occurrence of orbits in which one weakly bound electron performs a long Coulombic oscillation between two encounters. As a comparison with the analogous results for the model system (Eq. (12)) shows, it is not the long range of the Coulomb potential but the long time a trajectory can spend in the integrable region of phase space, which is responsible for the slower decay of the survival probability as function of time.

6 References

1. Friedrich H, Wintgen D (1989) Phys Reports 183: 37
2. Hasegawa H, Robnik M, Wunner G (1989) Prog Theor Phys suppl 98: 198
3. Friedrich H (1990) in: Atoms in Strong Fields, CA Nicolaides et al. eds, Plenum Press, New York: 247
4. Dando PA, Monteiro TS, Delande D, Taylor KT (1995) Phys Rev Lett 74: 1099
5. Hüpper B, Main J, Wunner G (1995) Phys Rev Lett 74: 2650
6. Schmelcher P, Cederbaum LS (1992) Z Phys D 24: 311
7. Schmelcher P, Cederbaum LS (1993) Phys Rev A 47: 2634
8. Schmelcher P, Cederbaum LS (1995) Phys Rev Lett 74: 662
9. Schmelcher P, Cederbaum LS (1996) contribution to this volume
10. Bayfield JE, Koch PM (1974) Phys Rev Lett 33: 258
11. Jensen RV (1984) Phys Rev A 30: 386
12. Leopold JG, Richards D (1985) J Phys B 18: 3369
13. Blümel R, Smilansky U (1987) Z Phys D 6: 83
14. van Leeuwen KAH, v. Oppen G, Renwick S, Bowlin JB, Koch PM, Jensen RV, Rath O, Richards D, Leopold JG (1985) Phys Rev Lett 55: 2231
15. Richards D (1989) in: Classical Dynamics in Atomic and Molecular Physics, Grozdanov T et al. (eds) World Scientific, Singapore: 269
16. Gebarowski R, Zakrzewski J (1995) Phys Rev A 51: 1508
17. Benvenuto F, Casati G, Shepelyansky DL (1993) Phys Rev A 47: R786
18. Benvenuto F, Casati G, Shepelyansky DL (1994) Z Phys B 94: 481
19. Casati G, Guarneri I, Mantica G (1994) Phys Rev A 50: 5018
20. Wiedemann H, Mostowski J, Haake F (1994) Phys Rev A 49: 1171
21. Su Q, Eberly H, Javanainen J (1990) Phys Rev Lett 64: 1990
22. Pont M, Shakeshaft R (1991) Phys Rev A 44: R4410
23. Eberly JH, Kulander KC (1993) Science 262: 1229
24. Kulander KC (1996): contribution to this volume
25. Grobe R, Law CK (1991) Phys Rev A 44: R4114
26. Tél T, Ott E (eds) (1993) Chaos Focus Issue on "Chaotic Scattering", Chaos 3: 417–706
27. Wiesenfeld L (1991) Proc Adriatico Research Conf on Quantum Chaos, Cerdeira H et al. (eds) World Scientific, Singapore: 371
28. Wiesenfeld L (1992) J Phys B (1992) 25: 4373
29. Eckhardt B (1987) J Phys A 20: 5971
30. Gaspard P, Rice S (1990) J Chem Phys 90: 2225
31. Gaspard P, Rice S (1990) J Chem Phys 90: 2242
32. Ott E, Tél T (1993) Chaos 3: 417
33. Blümel R, Reinhardt WP (1993) Directions in chaos. Feng DH and Yuan YM (eds) World Scientific, Singapore: Vol 4, p. 245
34. Smilansky U (1992) Chaos and Quantum Physics, Giannoni MJ et al. (eds) North Holland, Amsterdam: 371
35. Lai Y-C, Blümel R, Ott E, Grebogi C (1992) Phys Rev Lett 68: 3491
36. Eckhardt B (1993) Chaos 3: 613
37. Wirzba A (1993) Nucl Phys A560: 136
38. Vattay G, Wirzba A, Rosenqvist P (1994) Phys Rev Lett 73: 2304
39. Gruckenheimer J, Holmes P (1983) Nonlinear Oscillations, Dynamical Systems and Bifurcations of Vector Fields, Springer-Verlag, New York
40. Meiss JD, Ott E (1985) Phys Rev Lett 55: 2741
41. Lau Y-T, Finn JM, Ott E (1991) Phys Rev Lett 66: 978
42. Wiesenfeld L (1991/92) in: Irregular Atomic Systems and Quantum Chaos, Gay JC (ed) Gordon and Breach (1992) 335, *also published in*: Comm At Mol Phys 25 (1991) 335
43. Lai Y-C, Grebogi C, Blümel R, Ding M (1992) Phys Rev A 45: 8284
44. Hillermeier C (1991) doctoral thesis, University of Munich (unpublished)
45. Hillermeier C, Blümel R, Smilansky U (1992) Phys Rev A 45: 3486
46. Gebarowski R, Zakrzewski J (1994) Phys Rev A 50: 4408
47. Gu Y, Yuan J-M (1993) Phys Rev A 47: R2442

48. Yamamoto T, Kaneko K (1993) Phys Rev Lett 70: 1928
49. Richter K, Wintgen D (1990) Phys Rev Lett 65: 1965
50. Richter K, Wintgen D (1991) J Phys B 24: L565
51. Ezra G, Richter K, Tanner G, Wintgen D (1991) J Phys B 24: L413
52. Wintgen D, Richter K, Tanner G (1992) Chaos 2: 19
53. Rost J-M (1994) Phys Rev Lett 72: 1998
54. Rost J-M (1994) J Phys B 27: 5923
55. Draeger M, Handke G, Ihra W, Friedrich H (1994) Phys Rev A 50: 3793
56. Handke G, Draeger M, Ihra W, Friedrich H (1993) Phys Rev A 48: 3699
57. Ihra W, Draeger M, Handke G, Friedrich H (1995) Phys Rev A 52: 3752
58. Handke G, Draeger M, Friedrich H (1993) Physica A 197: 113
59. Handke G (1994) Phys Rev A 50: R3651
60. a Dahlqvist P (1992) Chaos 2: 43
    b Dahlqvist P (1992) J Phys A 25: 6265
61. Kulander KC, Schafer KJ, Krause JL (1991) Phys Rev Lett 66 2601
62. Beeker A, Eckelt P (1993) Chaos 3: 487

Microwave Multiphoton Excitation and Ionization

T. F. Gallagher

Department of Physics, University of Virginia, Charlottesville, VA 22901, USA

Using Rydberg atoms and microwave fields it has been possible to observe virtually all one electron strong field phenomena. The attraction of these experiments is that they can be more controlled than most laser experiments, with the result that more quantitative information can be extracted. The insights gained from these experiments can be profitably transferred to optical experiments. To demonstrate the latter point we demonstrate that apparently non-resonant microwave ionization, in fact, occurs by resonant transitions through intermediate states. These experiments demonstrated clearly the power of Floquet analysis of such processes, and the ideas were subsequently applied to the analogous problem of laser multiphoton ionization.

1 Introduction 126

2 Experimental Methods 127

3 Microwave Ionization 129

4 Microwave Transitions in a Two Level System 133

4.1 Floquet Description of a Two Level System 137

5 Dynamic Effects in Pulses 140

6 Ionization Revisited 143

7 Conclusion 146

8 References 147

1 Introduction

It is now possible to reach intensities of 10^{20} W/cm^2 at the focus of a laser beam, producing an optical electric field of 2×10^{11} V/cm [1]. This field is almost a factor of one hundred times larger than the atomic unit of field, 5.14×10^9 V/cm, the field experienced by the electron in the ground state of hydrogen. Furthermore, it is far larger than the field required, classically, to ionize the hydrogen atom, $E = 1/16$ a.u., 3.2×10^8 V/cm, which corresponds to an intensity of 2.5×10^{14} W/cm^2.

It is evident that even for intensities in the range of 10^{12}–10^{13} W/cm^2 the interaction with the laser field is not a minor perturbation to the atom. Since lasers which can be focused to this intensity are becoming rather common, understanding atom-strong field interactions is a problem of growing practical importance. Unfortunately, it is difficult to study atom-strong field interactions with lasers and ground state atoms in a completely quantitative way. The problems arise from the facts that the interactions are generally nonlinear in the laser field and that the laser beam is usually spatially focussed and temporally pulsed, leading to 100% spatial and temporal variations in the intensity over the exposed atoms. As a result, the observed signals are generally integrals of the signals produced over the spatial volume of the focus and the duration of the pulse, and the interpretation of the results is rather model dependent.

An alternative, and quantitative, method of studying atoms in strong fields is to use Rydberg atoms and microwave fields, an approach which was pioneered by Bayfield and Koch [2]. In a Rydberg atom, an atom in a state of high principal quantum number n, the energy in atomic units is given by $W = -1/2n^2$. The atomic unit of energy is twice the ionization potential of hydrogen. Unless otherwise specified atomic units will be used. The Rydberg states are weakly bound and closely spaced, with a spacing $\Delta W = 1/n^3$ between adjacent Rydberg states. Rydberg atoms also have large orbits. The expectation value of the radial position r is $\langle r \rangle \cong (3/2)\, n^2$ for $n \gg \ell$ where ℓ is the orbital angular momentum quantum number [3, 4]. The electric dipole matrix elements between states of the same n are also of this size. The atomic units of length and charge are the Bohr radius, $a_0 = 0.53$ Å, and the electron's charge, e.

Since the electron in a Rydberg atom is both weakly bound and in such a large orbit it is easily influenced by external fields. For example, the atomic field at $\langle r \rangle$ is $E = 1/\langle r \rangle = 4/9n^4$, and the classical field for ionization is given by $E = 1/16n^4$. It is useful to evaluate the relevant fields and frequencies for a Rydberg state of $n = 20$. The energy is $W = -1/800 = 274$ cm^{-1}, and the $n = 20$–21 spacing is $\Delta W = 1/8000 = 27.4$ cm^{-1}. The classical field for ionization $E = 10^{-4}/256 = 2.01$ kV/cm. Finally, the $\Delta n = 0$, $\Delta \ell = 1$ electric dipole moments are 600 ea_0.

If we are to study strong field interactions using Rydberg states of $n = 20$ it is apparent that the strongest field required is the classical ionization field, 2 kV/cm. In addition, we must work at frequencies far below the 27 cm^{-1}

(810 GHz) separation of the adjacent n states. Both requirements are easily met using microwave frequencies below 30 GHz, and many experiments have been done at frequencies of approximately 10 GHz. There are two advantages of using Rydberg atoms and microwaves. First, it is possible to make Rydberg atom samples with spatial dimensions of 1 mm, a length far smaller than the 3 cm wavelength at 10 GHz. Thus, all the Rydberg atoms can be exposed to the same microwave field, and the spatial averaging of laser experiments avoided. In addition, the microwave field can be applied either continuously or in a pulse with a controlled temporal dependence so that the effects of the temporal variation can be studied quantitatively.

Rydberg atoms and microwave fields constitute an ideal system for the study of atom-strong field effects, and they have been used to explore the entire range of one electron phenomena [5]. Here we focus on an illustrative example, which has a clear parallel in laser experiments, a series of experiments which show that apparently non-resonant microwave ionization of nonhydronic atoms proceeds via a sequence of resonant microwave multiphoton transitions and that this process can be understood quantitatively using a Floquet, or dressed state approach.

In the following section the experimental approach is briefly described. The initial observations of microwave ionization and the completely non-resonant picture initially used to describe it are then presented. Then microwave multiphoton transitions in a two level system analogous to the rate limiting step of microwave ionization are described both experimentally and theoretically. Experiments on this two level system with well controlled pulses of microwaves to show the applicability of an adiabatic Floquet theory to pulses are then described. We finally return to microwave ionization to see evidence for the resonant nature of the process.

2 Experimental Methods

There are two approaches to studying Rydberg atoms in microwave fields. The first is to use fast beams of H and He produced by charge exchange. In the initial experiments the Rydberg atoms were produced directly by charge exchange [2], but in later experiments atoms in low lying states have been excited to specific high lying Rydberg states with a CO_2 laser [6]. In these experiments the atoms pass through a waveguide or microwave cavity, and the transit time of the Rydberg atom through the microwave structure defines the interaction time, which is typically ~ 100 ns. The final state analysis is carried out downstream.

The second approach, used with alkali and alkaline earth atoms, is to use a thermal beam which moves slowly, ~ 1 mm/μs. With this speed most of the Rydberg atoms would decay before traversing the microwave structure if they were excited upstream of it, and the only practical alternative is to do everything inside the microwave structure in a temporal sequence [7]. The atoms are excited to Rydberg states by dye laser pulses of 5 ns duration, are exposed to the microwaves to induce bound-bound transitions or ionization, and are then exposed to a field pulse to ionize selectively Rydberg atoms not ionized and drive ions or electrons produced out of the microwave structure to a detector.

The most commonly used microwave structures are closed cavities, and the one used by Pillet et al. for microwave ionization of Na is a good illustrative example [7, 8]. It is shown in Fig. 1. It is a piece of WR 90 waveguide 20 cm long closed at both ends to form a resonant cavity. As shown, the atomic beam enters the cavity from one side and the exciting dye laser beams from the opposite side. The waveguide contains a septum, which allows the application of a static field parallel to the microwave field and a pulsed field to ionize the Rydberg atoms and drive the resulting ions out of the cavity. Note that there is a hole in the center of the top of the cavity for extraction of the ions, and only atoms directly under the hole contribute to the detected signal. The hole diameter of 1 mm is much smaller than the waveguide dimensions and the typical time between excitation and the field pulse is 1 μs so the signal comes from atoms which have experienced fields which differ by at most a few percent.

The cavity shown in Fig. 1 is operated on the TE_{10n} modes with n odd. In the TE_{10n} modes the electric field is vertical, and the septum and the pumping slots have no effect. The odd n modes have an E field antinode at the center of the cavity and can be fed by a probe in the center of the cavity, or a quarter wave from the end as shown in Fig. 1. Typically the quality factor, Q, is 2000, so the

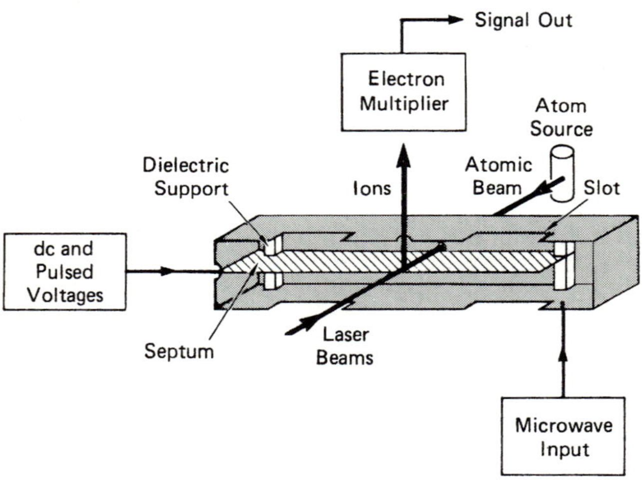

Fig. 1. Major components of a thermal atomic-beam apparatus for microwave ionization experiments, the atomic source, the microwave cavity, and the electron multiplier. The microwave cavity is shown sliced in half. The copper septum bisects the height of the cavity. Two holes of diameter 1.3 mm are drilled in the side walls to admit the collinear laser and Na atomic beams, and a 1-mm hole in the top of the cavity allows Na^+ resulting from a field ionization of Na to be extracted. Note the slots for pumping (from [8])

circulating power in the cavity is roughly 2000 times the input power, and with $Q = 2300$ at an operating frequency of 15 GHz 1 W of input power to the cavity shown in Fig. 1 produces a microwave field amplitude of 222 V/cm. While using a high Q cavity allows the use of low power sources, it precludes experiments requiring short microwave pulses, for the time constant τ for filling and decay of the cavity is $Q/2\pi f$, 25 ns for the Q and frequency given above. However, if the blank ends on the waveguide are replaced by waveguide to coaxial adapters, the waveguide can be used as a travelling wave structure and short, ~1 ns, pulses can be propagated through it [9]. The price is a reduced microwave field for a given input power.

The second kind of cavity which has been used is Fabry–Perot cavity [10]. These cavities are completely tunable and have the advantage that they are open, allowing far better access to the atoms than the closed cavity shown in Fig. 1. The improved access was critical for measurements of angular distributions of the electrons ejected in microwave ionization [11] Their cylindrical symmetry is useful for measurements involving circularly polarized microwaves, but a closed cylindrically symmetric cavity is equally good [12].

3 Microwave Ionization

The initial measurements of microwave ionization of Na were done with the cavity shown in Fig. 1 with 15 GHz microwave pulses of duration 200 ns–1 μs. The atoms were first excited to a Rydberg state by the laser pulses, then the microwave pulse was injected into the cavity. Finally, a high voltage pulse was applied to the septum to field ionize atoms left in the initial state and expel ions produced both by field ionization and microwave ionization from the cavity. With a judicious choice of pulsed field amplitude the two signals are temporally separated. The amplitude of the microwave pulse is varied over the course of many laser shots to measure the microwave ionization as a function of the microwave field. In Fig. 2 we show microwave ionization and field ionization signals from the Na 20s state. As can be seen in Fig. 2 the two signals are complementary, and we shall term the field at which 50% ionization occurs the ionization threshold field. It is also worth noting that close examination of the field ionization signal when the microwave ionization is incomplete reveals that the atoms are in the 20s or the adjacent $n = 19$ high ℓ states. They are not left in higher n states, which is true even for pulses as short as 5 ns [9].

When the measurements shown in Fig. 2 are repeated for many n states we find the n dependence of threshold fields shown in Fig. 3 for states of $|m| = 0$ and 1 and $|m| = 2$, m being the azimuthal orbital angular momentum quantum number. The $|m| = 0, 1$ states have ionization fields at $E \approx 1/3n^5$, while the $|m| = 2$ states have ionization fields of $E = 1/9n^4$. The $|m| = 2$ states are composed of $\ell \geqslant 2$ states all of which have quantum defects less than 0.015 and are

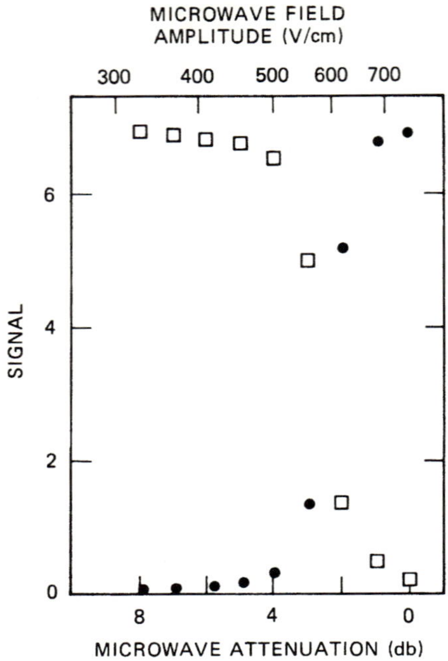

Fig. 2. Field-ionization signal (□) and 15 GHz microwave-ionization signal (●) for the Na 20s state showing the ionization threshold as both the disapperance of the field-ionization signal and the appearance of the microwave ionization signal with increasing microwave power (decreasing attenuation). For convenience the microwave-field amplitude is also given (from [8])

effectively hydrogenic. For the values of n shown in Fig. 3 the $|m| = 2$ states are ionized from the n state initially populated. Ionization proceeds via the lowest energy Stark state, hence the $1/9n^4$ dependence of the ionization field. While the microwave ionization of hydrogen has been intensely studied as an example of classical chaos [13, 14], our primary interest is the nonhydrogenic $|m| = 0$ and 1 states, which have ionization fields of $E \approx 1/3n^5$. The data of Fig. 3 were taken with 15 GHz microwave fields, but there is nothing special about 15 GHz. Similar $E \cong 1/3n^5$ behavior is seen in Na at frequencies as low as 670 MHz [15], and other atoms show similar behavior [6, 16].

The initial explanation of how the ionization occurred and why it had an $E \approx 1/3n^5$ scaling was based on a quasistatic field picture [6–8], which is easily understood by starting with the Stark effect in a static field. In an electric field the Stark shifts of hydrogenic $m = 0$ states are given by [3, 4]

$$W_s = \tfrac{3}{2} n(n_1 - n_2) \tag{1}$$

where the difference of the parabolic quantum numbers, $n_1 - n_2$, ranges from $-n + 1$ to $n - 1$ in steps of two. The extreme Stark states have shifts of

$$W_s = \pm \tfrac{3}{2} n(n - 1). \tag{2}$$

The energy spacing between the n and $n + 1$ levels is given by $1/n^3$, so the extreme blue n Stark state and the extreme red $n + 1$ Stark state cross at

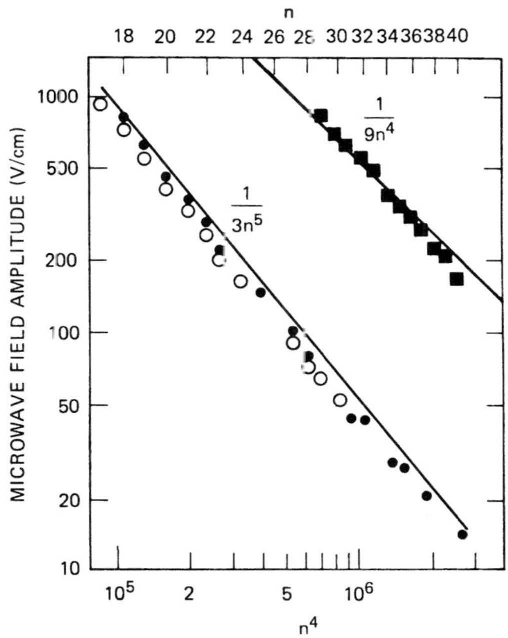

Fig. 3. 15 GHz microwave ionization fields for the $|m| \leqslant 1$ components of the Na $(n + 1)s$ (○) and nd (●) states. Ionization fields for the $|m| = 2$ components of the nd states (■) (from [8])

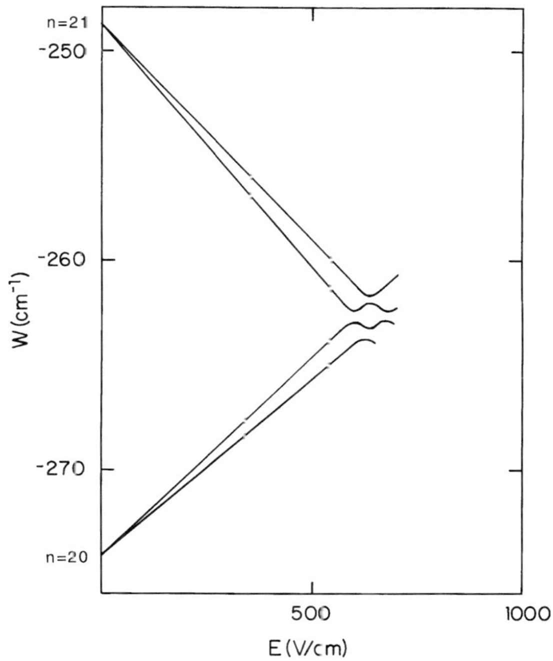

Fig. 4. Energy levels of the extreme $n = 20$ and $n = 21$, $m = 0$ states showing the avoided crossings of the blue $n = 20$ states and the red $n = 21$ states at fields above $E = 700$ V/cm

$E = 1/3n^5$. In the hydrogen atom these two levels cross, but in any other atom they are coupled by the finite sized ionic core and have an avoided crossing at $E = 1/3n^5$, as shown by Fig. 4. Consider the ionization of a Na atom initially excited to the 20d $m = 0$ state. As soon as the microwave pulse is turned on the $n = 20$ states are better described as Stark states, and the initial 20d population is rapidly spread over all the $n = 20$, $m = 0$ Stark states. When the microwave field rises to $E = 1/3n^5$ those atoms in the bluest Stark state are brought to the avoided crossing with the reddest $n = 21$ state and can make a Landau–Zener transition to it. Once the field is high enough that the $n = 20 \rightarrow n = 21$ transition occurs, it is far above the threshold for the transitions through higher lying states, they are made rapidly, and ionization occurs when it is classically allowed, as shown by Fig. 5.

For the blue $n \rightarrow$ red $n + 1$ transition probabilities on a single cycle to exceed 10% the microwave frequency must be of the same order of magnitude as the

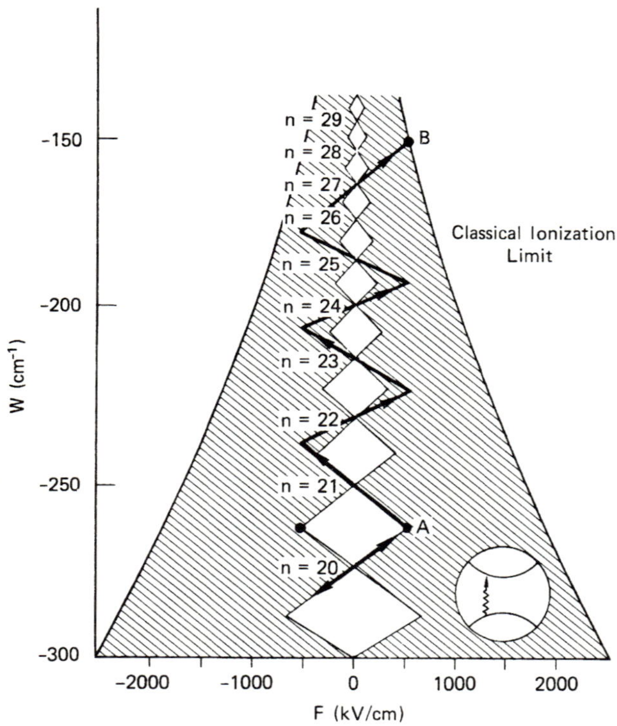

Fig. 5. Level diagram showing the mechanism by which an $n = 20$ atom is ionized by a microwave field of 700 V/cm amplitude. An atom initially excited to the 20d state is brought to the point at which the n-20 and 21 Stark manifolds intersect. At this point the atom makes a Landau–Zener transition, as shown by the *inset*, to the $n = 21$ Stark manifold and on subsequent microwave cycles make further upward transitions as shown by the *bold arrow*. The process terminates when the atom reaches a sufficiently high n state that the microwave field itselt ionizes the atom. This occurs at point B in this example (from [8])

$n \rightarrow n+1$ avoided crossing size and E must be approximately $1/3n^5$. For the $m = 0$ and 1 states the avoided crossings are large while for the hydrogenic $m = 2$ states they are orders of magnitude smaller, and ionization cannot occur by this Landau–Zener mechanism. Since there are many cycles in the microwave pulse the transition probability need not be high on a single cycle, and the required microwave field is slightly less than $E = 1/3n^5$, but still not quite as low as the measured ionization fields of Fig. 3. In sum, the single cycle Landau–Zener description provides the correct n dependence but predicts a field which is slightly too high. As we shall see, this discrepancy is due to the neglect of coherence over successive microwave cycles.

4 Microwave Transitions in a Two Level System

In our picture of microwave ionization the n dependence of the ionization fields comes from the rate limiting step between the bluest n and reddest $n+1$ Stark states. It would be most desirable to study this two level system in detail, but in Na this pair of Stark levels is almost hopelessly enmeshed in all the other levels. In K, however, there is an analogous pair of levels which is experimentally much more attractive [17, 18]. The K energy levels are shown in Fig. 6. All are $m = 0$ levels, and we are interested in the 18s level and the Stark level labelled (16, 3). We label the Stark states as (n, k) where n is the principal quantum number and k is the zero field ℓ state to which the Stark state is adiabatically connected. As shown in Fig. 6, the $(16, k)$ Stark states have very nearly linear Stark shifts and the 18s state has only a very small second order Stark shift, which is barely visible on the scale of Fig. 6. The 18s and (16,3) states have an avoided crossing at a field of 753 V/cm due to the coupling produced by the finite size of the K^+ core [19].

The pair of levels 21s – (16, 3) is exactly analogous to the extreme blue and red Na Stark states of n and $n+1$. The fact that only one has a permanent dipole moment is of no consequence; it is only the difference in the permanent moments which is significant. Based on the single cycle Landau–Zener description of microwave ionization we expect that if atoms in the 18s state are exposed to a microwave field of amplitude equal to the crossing field, $E_C = 753$ V/cm, they would make transitions to the (16, 3) state. On the other hand, if a static field is present as well as the microwave field it should be possible to see resonant microwave multiphoton transitions between these two bound states, and seeing the connection between these processes is part of our objective.

Using two 5 ns laser pulses atoms were excited to the K ns state of $18 \leqslant n \leqslant 23$ via the intermediate $4p$ state [17, 18]. The atoms were excited in zero static field to the $(n+2)s$ state and exposed to a 1 µs 9.278 GHz microwave pulse followed by a field ionization pulse set to ionize atoms in the $(n, 3)$ or higher lying states only. The field ionization signal was recorded as the

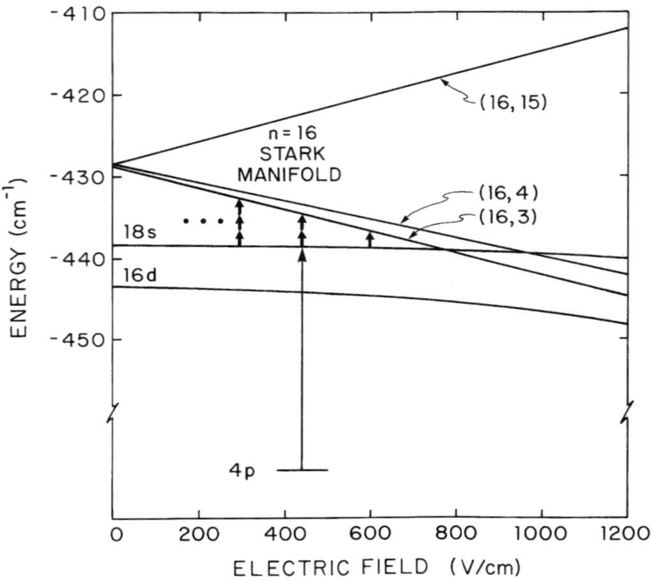

Fig. 6. Relevant energy levels of K near the $n = 16$ Stark manifold. The Stark manifold levels are labeled (n, k), where k is the value of ℓ to which the Stark state adiabatically connects at zero field. Only the lowest two and highest energy-manifold states are shown. The laser excitation to the 18s state is shown by the *long vertical arrow*. The 18$s \rightarrow$ (16, 3) multiphoton rf transitions are represented by the *bold arrows*. Note that these transitions are evenly spaced in static field, and that transitions requiring more photons occur at progressively lower static fields. For clarity, the rf photon energy shown in the figure is approximately five times its actual energy (from [17])

microwave amplitude was slowly varied over the course of many microwave pulses. In all cases but one a sharp threshold field was observed, at a field near, but always in excess of, the crossing field E_C. For the 19s state for which the anticrossing field is 546 V/cm, the microwave threshold field was 695 V/cm, but there was also a small signal at a microwave field of 515 V/cm. This behavior is qualitatively similar to that expected from the ionization experiments, in that the required microwave fields are close to the avoided crossing fields. However, there are two important differences. First, the measured fields are larger than expected, whereas in the ionization measurements they were smaller than expected. Second, the small prethreshold signal observed for the 19s – (17, 3) transition is completely new.

As shown by Fig. 6, since the $(n + 2)s$ and $(n, 3)$ states are both bound states it should be possible to observe resonant transitions between them. To observe the resonant transitions the K atoms were excited to the $(n + 2)s$ state by pulsed dye lasers in the presence of a static field E_s. The atoms were then exposed to a 1 μs microwave pulse followed by a field pulse to selectively ionize those atoms in the $(n, 3)$ state, or any higher lying state. The field ionization signal was detected as the static field was slowly scanned over many shots of the laser.

When the static field brought the two states into multiphoton resonance a sharp increase in the detected signal was observed, as shown in Fig. 7, which is a sequence of field scans showing the K $18s-(16,\ell)$ transitions with different microwave field amplitudes. In Fig. 7 the sequence of N photon $18s-(16,3)$ transitions, separated by 28 V/cm is quite apparent. At higher microwave fields transitions to other $(16, k)$ states are also present, cluttering the spectrum. From the data of Fig. 7 it is evident that the highest number of photons, N, which can be absorbed increases with the microwave field. In fact, it increases approximately linearly with the microwave field, and a linear fit of N to the microwave field yields

$$N = 0.063\,E_{\mathrm{mw}}(\mathrm{V/cm}) + 1.6. \tag{3}$$

A slightly different way of expressing the same notion is to express the required microwave field in terms of how far the N photon resonance is displaced from the anticrossing, which is the zero photon resonance. In these terms the microwave field $E_{\mathrm{mw},N}$ required to drive the N photon transition is given by

$$E_{\mathrm{mw},N} = 0.7(E_C - E_{S,N}) \tag{4}$$

where $E_{S,N}$ is the static field at which the N photon resonance is found. Finally, at low microwave fields the locations of the resonances do not change with microwave field, but at high microwave fields the resonances move slightly to higher static fields due to the second order AC Stark shift of the 18s state to lower energy. Since the Stark shift of the 18s state is due to its electric dipole couplings to p states, which are energetically far away compared to the microwave frequency, the shift due to the microwave field is the same as that of a static field of magnitude equal to the RMS value of the microwave field [18].

The first way we have examined the $(n+2)s \rightarrow (n,3)$ transitions, by varying the microwave field amplitude, corresponds to how we measured the microwave ionization threshold fields, a process explained by a single cycle Landau–Zener description. As noted above, the same $(n+2)s \rightarrow (n,3)$ transitions can also be observed as resonant transitions, raising the question of how the resonant transitions are related to the Landau–Zener description. Resonant transitions are simply the result of the coherence between Landau–Zener transitions on successive cycles. This notion is easily understood by examining Fig. 8. Imagine that atoms initially in the $(n+2)s$ state in a static field of E_S are exposed to a microwave pulse of amplitude high enough just to reach the avoided crossing, so that on the first cycle the wavefunction acquires a small $(n, 3)$ amplitude. The amplitude from the second cycle will add constructively if the accumulated phase difference between the two states over the field cycle is $2\pi N$, where N is an integer. This phase requirement is equivalent to requiring that the two states lie N photons apart at E_S. If there are only two cycles this situation is exactly analogous to the interference pattern in Young's two slit experiment, and we expect a sinusoidal variation of the transition probability with frequency. If there are more cycles the interference pattern resembles the interference pattern

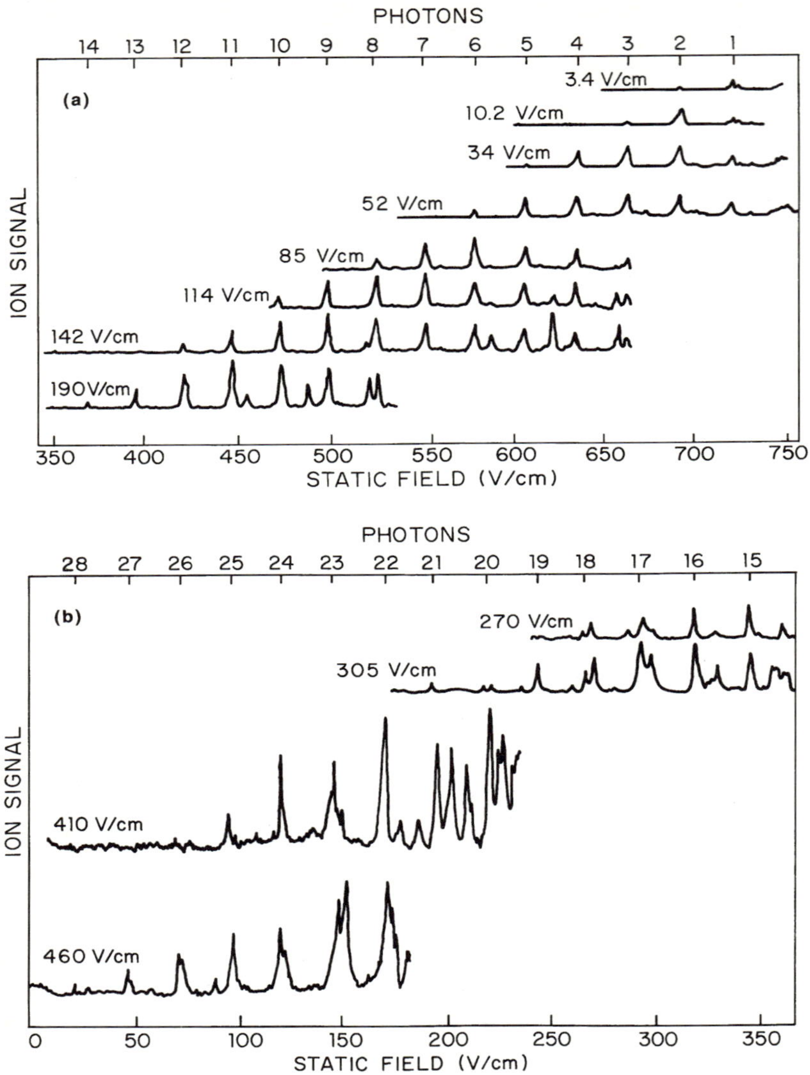

Fig. 7. **a** K $18s \to (16, 3)$ 1- to photon transitions observed as the static field is scanned from 350–750 V/cm for the 10.353.-GHz microwave fields indicated above each trace (3.4 to 190 V/cm). The regularity of the progression is quite apparent. Note the extra resonances in the 142 V/cm microwave field trace. These are due to $18s$ (16, 4) transitions. **b** $18s \to (16, 3)$ 15- to 28- photon transitions observerd as the static field is scanned from 0 to 350 V/cm. Note the congestion of the 410-V/cm trace at static fields above ~200 V/cm, due to many overlapping ($18s \to 16$, k) transitions (from [17])

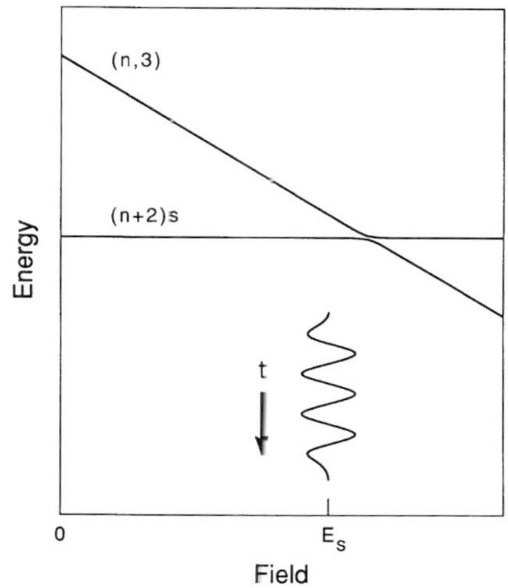

Fig. 8. Schematic picture of a multicycle Landau–Zener transition. In combined static and microwave fields the oscillating field brings the atom in the $(n + 2)s$ state to the avoided crossing on successive cycles, and the transition amplitudes due to successive cycles add, leading to interference, or resonances (from [5])

from a transmission grating and should exhibit sharp resonances as a function of microwave frequency.

Having examined the $(n + 2)s - (n, 3)$ transitions as resonances we are now able to explain the apparently random fields required to drive the $(n + 3)s - (n, 3)$ transitions by a microwave field alone. Observation of the transition has two requirements, the levels must be resonant and the Rabi frequency must be adequate. Analyzing the data of Fig. 7 show that the Rabi frequency is adequate if $E_{mw} = 0.7\,(E_C - E_{S,N})$ for all the n states. In the absence of a static field the resonance condition is met when the AC Stark shift of the $(n + 3)s$ state brings it into resonance, which is random. Typically, the two conditions are only met simultaneously for a random microwave field amplitude larger than the anticrossing field, but for the 19s state in a 9 2789 GHz field the resonance condition is met for the 27 photon transition at $E_{mw} = 515$ V/cm, $E_{mw} \approx 0.9 E_C$, leading to the small resonant peak in the signal [18].

4.1 Floquet Description of a Two Level System

Although extending the Landau–Zener description to many cycles accurately describes the evolution from the non-resonant interaction with a single cycle to the resonant interaction with many cycles, this approach is not very convenient to use nor does it easily lead to any analytic predictions. Instead we use a Floquet approach [20–22]. We assume that we have two states, the $(n + 2)s$ state and the $(n, 3)$ Stark state. The effective Hamiltonian,

$$H = H_{atom} + H_s + H_{mw} + H_{core} \tag{5}$$

has four terms corresponding to the atom in zero field and the effects of a static field, a microwave field polarized in the same direction, and the ionic core coupling between the levels. If we assume that there is no microwave field and ignore the core coupling, in a static field the $(n + 2)s$ state has an energy given by

$$W_s = W_{0s} - \frac{\alpha}{2} E_s^2 \tag{6}$$

where W_{0s}, α and E_s are the zero field energy, the polarizability, and the static field. The Stark state has an energy given by

$$W_k = W_{0k} - kE_s \tag{7}$$

where W_{0k} and k are the zero field energy and the permanent dipole moment respectively. The $(n + 2)s$ and $(n, 3)$ states cross at E_C, which, ignoring the second order shift of the $(n + 2)s$ state, is given by $E_C = (W_{0k} - W_{0s})/k$. When we introduce the core coupling, which lifts the degeneracy at the crossing, these two states have an avoided crossing at E_C. The separation ω_0 at the avoided crossing is given by

$$\omega_0 = 2|\langle (n + 2)s| V |(n, 3)\rangle|. \tag{8}$$

Now let us consider the effect of adding a microwave field to the static field, again ignoring the core coupling. When a microwave field $E_{mw}\cos\omega t$ is added to the static field the $(n + 2)s$ state is further shifted by $-\alpha/4E_{mw}^2$. The $(n + 2)s$ state also acquires a small amount of p character, which we ignore. The effect of a microwave field on the $(n, 3)$ Stark state is more dramatic. The oscillating field modulates its energy, and the Stark state breaks into a carrier and sidebands in the same way a frequency modulated radio wave does, so that the wavefunction is given by [23]

$$\Psi_{(n,3)}(\bar{r}, t) = \Psi_{(n,3)}(\bar{r})\, e^{-i(W_0k - kE_s)t} \sum_{m=-\infty}^{\infty} J_m\left(\frac{kE_{mw}}{\omega}\right) e^{-im\omega t}. \tag{9}$$

The m-th sideband state is displaced from the carrier, or the $(n, 3)$ state in the absence of microwaves, by $m\omega$. Note that the spatial wavefunction of Eq. (9) is unchanged.

The carrier and sideband states cross the $(n + 2)s$ state. If we ignore the AC Stark shift of the $(n + 2)s$ state the intersection occurs at a static field $E_S = E_C - m\omega/k$. When the core coupling is taken into account, the crossings become avoided crossings, and the magnitude of the avoided crossing between the $(n + 2)s$ state and the m-th sideband of the $(n, 3)$ state is given by

$$\omega_m = 2|\langle (n + 2)s| V \left|(n, 3)\rangle J_m\left(\frac{kE_{mw}}{w}\right)\right|. \tag{10}$$

The avoided crossing of the sideband state is the purely static field avoided crossing multiplied by the amplitude of the m-th sideband state. The size of the anticrossing is the m photon Rabi frequency at resonance. For large arguments $J_m(x)$ drops to zero when $m > x$ [24], or when $m\omega > kE_{mw}$. Thus the Rabi frequency and the microwave field required to drive the m photon transition are proportional to m, as shown earlier by Eq. (3).

The earliest measurements, shown in Fig. 6, had large enough field inhomogeneities to make accurate measurements of Rabi frequencies impossible. However, with some refinements Gatzke et al. [25] were able to measure the Rabi frequencies directly. With a pulsed dye laser they excited the K atoms from the $4p$ state to the $21s$ state in microwave and static fields at the static fields of the m photon $21s-(19, 3)$ resonances. At these resonances the energy eigenstates ψ_+ and ψ_- are the linear superpositions $\psi_\pm = 1/\sqrt{2}(\psi_{21s} \pm \psi_{(19,3)})$, and the coherent superposition of the energy eigenstates corresponding to purely $21s$ is excited by the short laser pulse. The two $\psi_\pm$ energy eigenstates evolve at different rates in time and the wave function oscillates back and forth between the $21s$ and $(19, 3)$ in time at the m photon Rabi frequency. To detect this oscillation the states are abruptly detuned from resonance at a variable time after excitation and the atoms in the $(19, 3)$ state detected. The Rabi oscillations back and forth between $21s$ and 19,3 states are clearly seen in the signal, as shown in Fig. 9, which shows the Rabi oscillations for the 4 photon transition at several microwave field amplitudes.

The results of a series of measurements such as those shown in Fig. 9 are shown in Fig. 10. It is clear that the observed Rabi frequencies are in excellent agreement with the Bessel function expression of Eq. (10). In particular, it is interesting to note that the zeros in the Bessel function are reproduced in the experimental data. These data show clearly that the Floquet description presented above is indeed accurate for Rydberg atoms in continuous wave microwave fields.

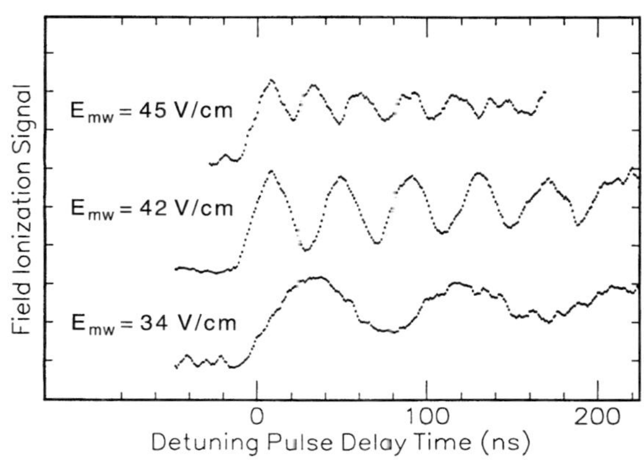

Fig. 9. Four photon Rabi oscillations between the K $21s$ and 19, 3 states at 9.1 GHz for microwave amplitudes of 34, 42, and 45 V/cm (from [25])

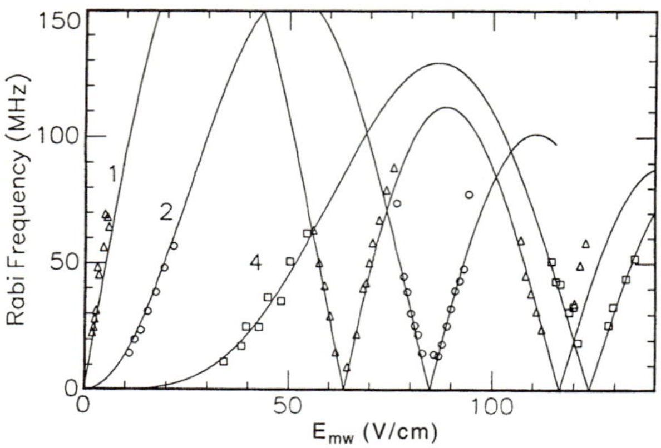

Fig. 10. K $21s - 19, 3$ Rabi frequencies for 1, 2, and 4 photon 9.1 GHz resonances as a function of the microwave field amplitude (from [25])

5 Dynamical Effects in Pulses

In the Floquet description given above it is apparent that the $(n, 3)$ state breaks into an infinite set of sideband states all of which are evenly spaced by the microwave angular frequency. In fact, the $(n + 3)s$ state also has sidebands, spaced by the microwave frequency ω. We ignored them since, in the sideband picture given above, only the carrier has appreciable amplitude. In our two level problem there are two sets of energies spaced by ω, and thus only two physically meaningful energies, the two Floquet energies, in any frequency interval [21,22]. If there are M states there are M Floquet energies.

For any fixed microwave and static fields there exists a well defined set of Floquet energies. As a consequence we can make plots of the Floquet energy levels as functions of the static field or the amplitude or frequency of the microwave field. No matter which variable we choose there are avoided level crossings at the multiphoton resonances, and how large they are depends mostly on the microwave field amplitude. As long as the field amplitude, frequency, or the static field change slowly compared to the microwave frequency we can determine the dynamics from the evolution of the Floquet states [26]. A particularly useful application is the response to a microwave pulse. As shown numerically by Breuer et al., as long as the pulse is longer than ten cycles we can ignore the rapid oscillation of the field itself and calculate the effect of the pulse using the evolution of the Floquet energy levels with the more slowly varying envelope of the pulse [26]. For example, if the initial state has an AC stark shift it can shift through multiphoton resonance with the final state on the rising and falling edges of the pulse, and the avoided crossings at the multiphoton resonance can be traversed adiabatically or diabatically depending on the speed of the traversal. In short, we have recast the atom-radiation pulse problem in the language of the theory of slow atomic collisions. A specific example of this process is the four photon 9.3 GHz transition from the K 21s state to the (19, 3) state [27]. At zero microwave power the 21s and (19, 3) Floquet state, the (19, 3) state minus 4 photons, cross at a static field of 236 V/cm. If the static field is slightly above this value, the 21s state can be shifted into resonance at a finite microwave field by its AC Stark shift. As shown by Fig. 11, at the resonance there is an avoided crossing of magnitude $\omega_4 = \omega_0 |J_4(kE_{\text{mw}}/\omega)|$ where ω_0 is the magnitude of the static field anticrossing at 304 V/cm. The microwave field required to bring the two states into resonance depends on the static field and the polarizability of the 21s state, and the microwave field at resonance determines the size of the avoided crossing as shown by Eq. (10). As seen in Fig. 11, different static fields lead to different avoided crossings.

In a pulse of microwaves atoms initially in the 21s state traverse the avoided crossing twice, if at all, on the rising and falling edges as shown in Fig. 11. Typically the crossings are traversed partially diabatically, leading to two pathways which interfere. Depending on the phase difference Φ, given by [27]

$$\Phi = \int_{t_-}^{t_2} (W_{(19,3)} - W_{21s})\,\mathrm{d}t, \tag{11}$$

the transition amplitude of the second traversal on the falling edge can add constructively or destructively to the amplitude of the first traversal, on the

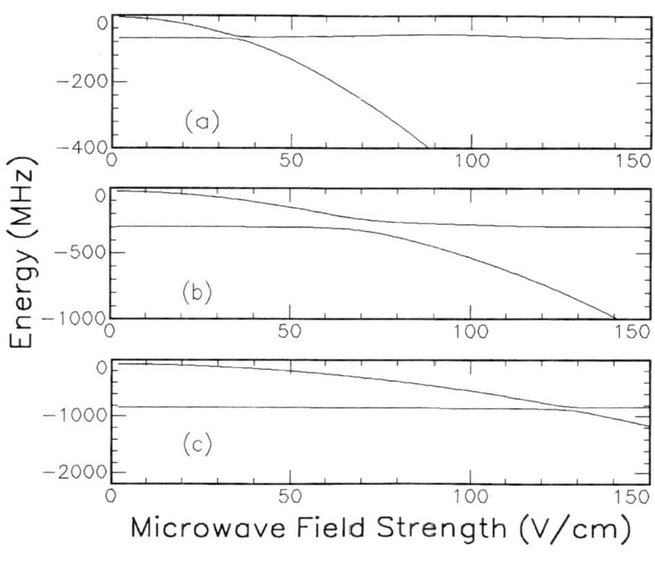

Fig. 11a–c. The Floquet energies W_s and W_k, corresponding to the 21s and 19, 3 states, as functions of the microwave field E_{mw} for the static fields E_s of: **a** 236.7 V/cm; **b** 237.2 V/cm; **c** 238.4 V/cm. W_k is displaced from (a) to (c) by the linear 19, 3 static Stark shift, and W_s is displaced primarily by the ac Stark shift. The avoided crossing is 12, 99, and 35 MHz in (a), (b) and (c) (from [27])

rising edge. If $\Phi = 2\pi N$ where N is an integer the two amplitudes add constructively, while if $\Phi = 2\pi(N + 1/2)$ the interference is destructive. The interference is similar to Young's two slit interference pattern, Ramsey fringes, and Stuckelberg oscillations.

These notions were verified experimentally by exciting K atoms to the 21s state in a static field, applying a short microwave pulse of fixed amplitude and detecting the atoms which had undergone transitions to the (19, 3) state [25]. Analogous experiments were done with He [28]. Slowly, varying the static field over many laser shots led to the oscillatory signals shown in Fig. 12a. The signals of Fig. 12a were obtained with 8 ns pulses and a peak field amplitude of 116 V/cm. Although it is not shown here, the oscillations become faster if the pulse is made longer or higher in amplitude, as expected. For $E_s < 236.5$ V/cm no signal is observed since the two states are never in resonance. From 236.5–237.5 V/cm there are clear oscillations due to the change in the phase Φ with the static field. As the static field is raised the amount of time the 21s state lies below the (19, 3) state decreases and Φ decreases accordingly. As expected, the oscillations in the signal become faster as the pulse is made longer or higher in amplitude. In Fig. 12b we show the signals obtained with an 8 ns pulse of 178 V/cm peak amplitude. The oscillatory structure seen at 236.5–237.5 V/cm is no longer resolved, but new features appear at higher static fields. At $E_S = 237.5$ V/cm the signal almost disappears, for then static field levels come into resonance at $E_{mw} = 104$ V/cm, which is at the peak of $J_4(kE_{mw}/\omega)$, where the 4 photon Rabi frequency is 125 MHz, as shown in Fig. 11b. This Rabi frequency is so large that the avoided crossing is traversed adiabatically, and all the atoms pass from the 21s state to the (19, 3) state and back to the 21s state during the pulse. The Rabi frequency is too large for the transition to occur, a somewhat counter intuitive result. As the static field is raised above $E_S = 237.5$ V/cm the signal increases since the Rabi frequency is decreasing. The signal disappears sharply at $E_S = 238.6$ V/cm, for at this value of the static

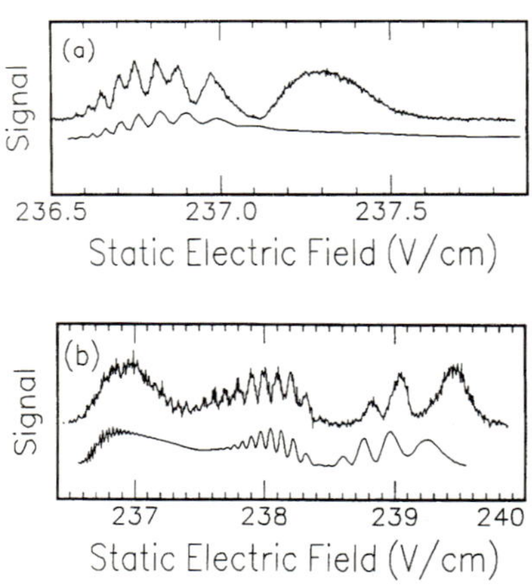

Fig. 12a, b. The (19, 3) signal as a function of the static tuning field for 8 ns pulses of two different peak amplitudes: **a** 116 V/cm; **b** 178 V/cm. In both cases the *arrows* show the static field at which the two states are brought into resonance at the peak of the pulse. The *light lines* are the results of direct numerical integration of the Schroedinger equation (from [27])

field the two states are brought into resonance at the microwave field $E_{mw} = 135$ V/cm, which produces the first zero of $J_4(\kappa E_{mw}/\omega)$. Beyond $E_S = 238.6$ V/cm the signal reappears. The light lines of Fig. 12 are simulations, based on numerical integration of the Schroedinger equation, which are in good agreement with the experimental signals, although some fine features of the simulations are not reproduced in the experimental data, presumably due to field inhomogeneities. In sum, the Floquet picture gives an excellent representation of the response to a pulse of microwaves.

6 Ionization Revisited

It is interesting to consider now the evidence in the microwave ionization data which support the assertion that it is a resonant process. There are three different kinds of observations which support this assertion, and they are outlined below.

First, the microwave ionization fields fall consistently below $E = 1/3n^5$. For example the data of Fig. 3 are fitted by $E = 1/3.7n^5$, a field which is only 80% of the crossing field. This field seems too low to account for the observed ionization rates if transition probabilities from individual cycles are added incoherently, but at 15 GHz it is not really clear if there is a discrepancy between the single cycle picture and the experiments or not. However, measurements of the microwave ionization of Na at frequencies as low as 670 MHz show an $E = 1/3n^5$ dependence [15]. This low a frequency is much smaller than the avoided crossing size, enormously reducing the transition probability on a single cycle, and there are so few cycles that adding incoherently the effects of successive cycles cannot lead to ionization [15]. The coherent addition of transition amplitudes on successive cycles must play an important role.

Second, the optical excitation spectra of the atoms in strong microwave fields shows that there is coherence in the microwave field and that the onset of ionization corresponds to the overlap of the sidebands of adjacent n states [29]. In these experiments Na atoms were excited from the $3p$ to Rydberg states in the presence of a 15 GHz field. The excitation spectra were obtained by scanning the wavelength of the exciting laser over the region encompassing $n = 25$ and 26 and detecting all the Rydberg atoms formed. As shown in Fig. 13, in zero microwave field only the s and d states are observed. In relatively small microwave fields, 25 V/cm, the doublet corresponding to the $26p$ state ± 1 microwave photon appears at a binding energy of -174 cm^{-1}. At higher fields the sideband states separated by 15 GHz, 1/2 cm^{-1} are plainly visible, showing that the atoms are in well defined Floquet energy states in the strong microwave field. It is most interesting to note that the sideband states are observed in the region where states exist in a static field, as shown by the light lines. Finally, note that the $n = 25$ and $n = 26$ sidebands begin to overlap at a microwave field of 150 V/cm, which coincides with the $n = 25$ threshold field for ionization.

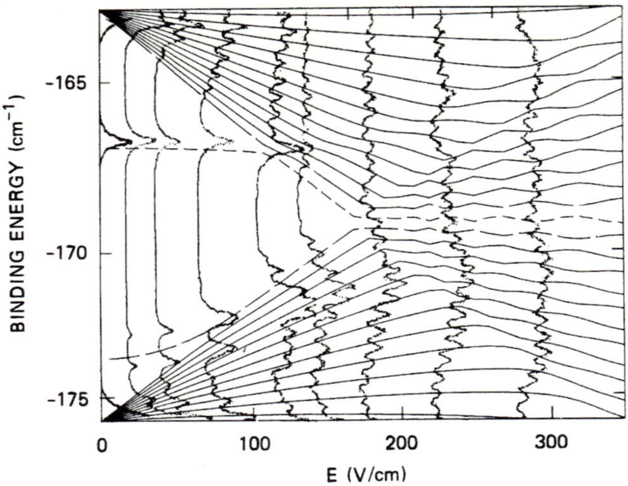

Fig. 13. Observed Na 3p $\rightarrow n = 25, 26$ spectra in varying strengths of microwave field overlaid on an energy-level diagram of the Na $|m| = 0$ states. The baseline of each spectrum is located at the amplitude of the microwave field. Note that only odd sidebands of the *p* states occur and that the predominant effect of higher microwave fields is to add more sidebands (from [29])

The third piece of evidence for resonant transitions in microwave ionization occurs in Li. In Li a small static field produces a dramatic effect on microwave ionization [30]. For example, with no static field the Li 42d state is ionized by a microwave field of $E = 200$ V/cm, a field which coincides with the hydrogenic $E_{mw} = 1/9n^4$ dependence of the microwave ionization field. When a static field of 1 V/cm, parallel to the microwave field, is added, the ionization exhibits a broad threshold centered at 20 V/cm, a field slightly in excess of $E = 1/3n^5 = 13$ V/cm. To make systematic measurements of the static field needed the microwave power was held fixed at several values and the static field swept while detecting the atoms not ionized by the microwaves. Typical results, for $n = 36$, are shown in Fig. 14. As shown, for 15.217 GHz microwave fields of 44, 70, and 140 V/cm for static fields larger than 1 V/cm there is a significant depletion of the atoms not ionized. Repeating the procedure leading to Fig. 14 for different *n* states and determining the half width of the observed peaks provides a measure of the *n* dependence of the static field required to produce the enhancement in microwave ionization. Changing the microwave frequency to 7.921 GHz yields the frequency dependence, and the required static field in atomic units is [28]

$$E_S = 0.1\omega/n^2. \tag{12}$$

Before describing the precise origin of the variation shown in Eq. (12) it is worth noting that the observation of the enhancement of ionization by such a small field is incompatible with a single cycle Landau–Zener description. In contrast, the $0.1\omega/n^2$ dependence of the static field required is easily explained with a resonance picture. As shown in Fig. 15, the sidebands of a given *n* state in

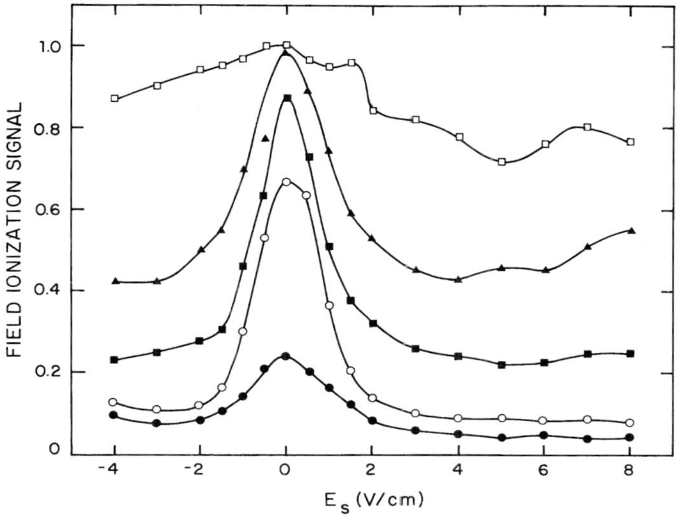

Fig. 14. Field-ionization signals from remaining atoms when the Li 36d state is initially excited in a variable static field with the 15.217-GHz microwave field amplitude varied as a parameter. The data are all normalized to the observed signals with $E_{mw} = 0$ at the same static field. $E_{mw} = 31$ (*open squares*), 44 (*triangles*), 70 (*filled squares*), 140 (*open circles*), and 307 V/cm (*filled circles*). Note that with $E_{mw} = 44$, 70, and 140 V/cm, there is a rapid decrease in the number of remaining atoms or, equivalently, a rapid increase in the microwave-ionization probability, at $\simeq 1$ V/cm, corresponding to the field required for quasicontinuum production (from [30])

a microwave field E_{mw} span the energy range from $-3/2n^2E_{mw}$ below to $-3/2n^2E_{mw}$ above the zero field energy, which is the same energy range as spanned by the Stark states of that n in a static field of the same size. Note that all Stark states of the same n have sidebands which are degenerate. The only difference between states with large and small Stark shifts is how many sidebands have appreciable amplitudes. As a rule there is a detuning Δ between the n and $n+1$ sidebands, and for Li this detuning is generally larger than the multiphoton Rabi frequency from the avoided crossing ω_0 between the n and $n+1$ levels in a static field. As a result, with no static field no $n \rightarrow n+1$ transitions occur. When a static field E_S is applied it separates the sidebands of different Stark states. The sidebands of the extreme Stark states are shifted by $-3/2n^2E_s$ and $-3/2n^2E_s$, while the Stark states with smaller shifts have less shift. The effect on the sideband spectrum is shown in Fig. 15c. As described above, the extreme sidebands for any n are from the extreme Stark states and are split into two components separated by $3n^2E_S$. The central sidebands come from all Stark states and have virtually all possible shifts, forming a continuous band of states. If $3n^2E_S = \omega$ or $E_S = \omega/3n^2$ the spectrum becomes quasi continuous for the central sidebands of each n state. A smaller static field will produce a quasi continuous spectra over a proportionally smaller region. Considering that both the n and $n+1$ states have been converted into quasi continuous

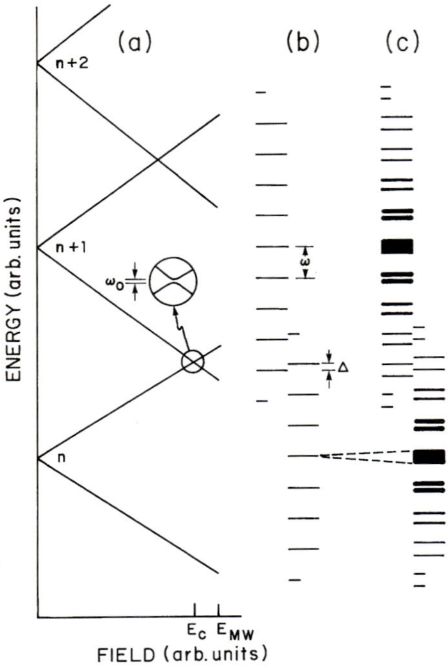

Fig. 15. a Static energy-level diagram showing the extreme $|m| = 0$ Stark states of then $n, n+1$, and $n+2$ Stark manifolds. The extreme n and $n+1$ levels have an avoided crossing at a field of $E_c = 1/3n^5$. As shown by the inset, the magnitude is ω_0. If a microwave field $E_{mw} \geqslant E_c$ is applied, atoms can make the Landau–Zener transition $n \rightarrow n+1$ at the avoided crossing, followed by successive analogous transitions, culminating in ionization. **b** Spectrum of sideband states produced by a microwave field of amplitude E_{mw} and angular frequency ω acting on the extreme n and $n+1$ Stark states. The sidebands from all Stark states of the same n are degenerate modulo ω. In general there is a detuning Δ between the n and $n+1$ sideband states. **c.** Spectrum of n and $n+1$ sidebands states upon the application of a small electric field which removes the degeneracy of the sideband states originating from different Stark states (from [30])

spectra and that the field of Eq. (12) corresponds to producing half the possible enhancement of the ionization, we feel that the resonance picture of Fig. 15 provides an excellent match to the data. This Li experiment shows clearly that microwave ionization is a resonant process, raising the question of why similar effects are not observed in Na, where adding a small static field simply lowers the microwave ionization field, as expected from a single cycle Landau–Zener picture [8]. In Na the $n \rightarrow n+1$ avoided crossings are larger, so that enough accidental resonances occur that adding the static field is unnecessary. Furthermore, the presence of two Na states, *s* and *p*, with large quantum defects leads to larger second order Stark shifts, making it possible that resonances encountered on the rising edge of the microwave pulse play a role.

7 Conclusion

Microwave ionization of nonhydrogenic atoms, first thought to be a non-resonant process, has, by a series of experiments, been shown to be a resonant process. There is a clear parallel with optical experiments, where what had been thought to be non-resonant ionization was shown to be resonantly enhanced by Stark shifts due to the optical field [31]. As in the microwave experiments, all the details were not immediately worked out [32, 33], but, finally, a dynamic Floquet description similar to the one outlined here provides a good description of the process [34, 35].

Acknowledgements. This work has been supported by the Air Force Office of Scientific Research. It represents the synthesis of the contributions of many, and it is a pleasure to acknowledge the contributions of P. Pillet, H.B van Linden van den Heuvell, J.L. Dexter, R.C. Stoneman, C.R. Mahon, M. Gatzke, and M.C. Baruch. Without their efforts microwave processes would not be as well understood as they are.

References

1. Svanberg S, Larsson J, Persson A, Wahlström CG (1994) Phys Scr 49: 187
2. Bayfield JE, Koch PM (1974) Phys Rev Lett 33: 258
3. Bethe HA, Salpeter EA (1957) Quantum mechanics of one two electron atoms. Academic Press, New York
4. Gallagher TF (1994) Rydberg atoms. Cambridge University Press, Cambridge
5. Gallagher TF (1992) In: Gavrila M (ed) Atoms in intense laser fields. Academic Press, Cambridge
6. Mariani DR, van de Water W, Koch PM, Bergeman T (1983) Phys Rev Lett 50: 1261
7. Pillet P, Smith WW, Kachru R, Tran NH, Gallagher TF (1988) Phys Rev Lett 50: 1042
8. Pillet P, van Linden van den Heuvell HB, Smith WW, Kachru R, Tran NH, Gallagher TF (1984) Phys Rev A 30: 280
9. Gatzke M, Baruch MC, Watkins RB, Gallagher TF (1994) Phys Rev 50: 2502
10. Fu P, Scholz TJ, Hettema JM, Gallagher TF (1990) Phys Rev Lett 64: 511
11. van de Water W, Yoakum S, van Leeuwen T, Sauer BE, Moorman L, Galvez EJ, Mariani DR, Koch PM (1990) Phys Rev A 42: 872
12. Cheng CH, Lee CY, Gallagher TF (1994) Phys Rev Lett 73: 3078
13. van Leeuwen KAH, Oppen GV, Rennick S, Bowlin JB, Koch PM, Jensen RV, Rath O, Richards D, Leopold JG (1985) Phys Rev Lett 55: 2231
14. Jensen RV, Susskind SM, Sanders MM (1991) Phys Rept 201 1
15. Mahon CR, Dexter JL, Pillet P, Gallagher TF (1991) Phys Rev A 44: 1859
16. Eichmann U, Dexter JL, Xu EY, Gallagher TF (1989) Z Phys D 11: 187
17. Bloomfield LA, Stoneman RC, Gallagher TF (1986) Phys Rev Lett 57: 2512
18. Stoneman RC, Thomson DS, Gallagher TF (1988) Phys Rev A 37: 1527
19. Stoneman RC, Janik GR, Gallagher TF (1986) Phys Rev A 34: 2952
20. Shirley J (1965) Phys Rev 138: 13979
21. Sambe H (1973) Phys Rev A 7: 2203
22. Christiansen-Dalsgaard B (1990) unpublished
23. Autler SH, Townes CH (1955) Phys Rev Lett 100: 703
24. Abramowitz M, Stegun IA (1964) Handbook of Mathematical Functions, Nat Bur Stand Appl Math Ser No 55, US GPO, Washington, DC
25. Gatzke MA, Baruch MC, Watkins RB, Gallagher TF (1993) Phys Rev A48: 4742
26. Breuer HP, Dietz K, Holthaus M (1988) Z Phys D 10: 13
27. Baruch MC, Gallagher TF (1992) Phys Rev Lett 68: 3515
28. Yoakum S, Sirko L, Koch PM (1992) Bull Am Phys Soc 37: 1105
29. van Linden van den Heuvell HP, Kachru R, Tran NH, Gallagher TF (1984) Phys Rev Lett 53: 1901
30. Pillet P, Mahon CR, Gallagher TF (1988) Phys Rev Lett 60: 21
31. Freeman RR, Bucksbaum PH, Milchberg H, Darack S, Schumacher D, Geusic ME (1987) Phys Rev Lett 59: 1092
32. deBoer MP, Muller HG (1992) Phys Rev Lett 68: 2747
33. Gibson GN, Freeman RR, McIlrath TJ (1992) Phys Rev Lett 69: 1904
34. Story JG, Duncan DI, Gallagher TF (1993) Phys Rev Lett 70: 3012
35. Vrijen RB, Hoogenraad JH, Noordam LD (1993) Phys Rev Lett 70: 3016

Time-Dependent Calculations of Electron and Photon Emission from an Atom in an Intense Laser Field

K. C. Kulander[1] and K. J. Schafer[2]

[1] Physics Directorate, Lawrence Livermore National Laboratory, Livermore, CA 94551
[2] Physics Department, Louisiana State University, Baton Rouge, LA 70803

A numerical method for solving the time-dependent Schrödinger equation for a one-electron atomic system in an intense, short-pulsed laser field is presented. An effective potential formalism is proposed and tested for representing the excitation of the valence electrons in rare gases. Results for ion production yields, photoelectron distributions and harmonic conversion are presented and compared to recent experimental results.

1 Introduction 150

2 Method and Model 151
- 2.1 Time-Dependent Theory of Multiphoton Processes 151
- 2.2 One-Electron Systems 152
- 2.3 Multi-Electron Systems 154
  - 2.3.1 Time-Dependent Hartree-Fock 154
  - 2.3.2 Single-Active-Electron Approximation 155
- 2.4 Calculation of Emission Spectra 159
  - 2.4.1 Photoemission Spectra 159
  - 2.4.2 Photoelectron Distributions 159

3 Results 160
- 3.1 Ionization 160
- 3.2 Population Transfer 162
  - 3.2.1 Rydberg Trapping 162
  - 3.2.2 Single State Trapping 164
- 3.3 Above Threshold Ionization 166
- 3.4 Harmonic Generation 168

4 Conclusion 170

5 References 171

Structure and Bonding, Vol. 86
Springer-Verlag Berlin Heidelberg 1997

1 Introduction

In recent years there have been significant advances in the technology of short pulse, high intensity lasers. Lasers with pulse lengths of 0.01–1 ps and wavelengths from 0.2–1 μm can be focused to produce intensities from 10^{12} to above 10^{18} W/cm^2. Some of these systems can emit of the order of a thousand pulses per second allowing very high precision measurement of the phenomena of interest. One major application has been to study the response of atoms and molecules to such intense, well characterized electromagnetic fields. These laser pulses are short enough that neutral atoms survive to experience intensities where traditional perturbation expansions, based on the field-free states of the system, fail to describe the dynamics of the system. Evidence for non-perturbative ionization can be found in the photoelectron energy spectra which contain numerous peaks separated by the photon energy. These peaks indicate a high probability of absorbing many more photons than the minimum required for ionization. This phenomenon, called above threshold ionization (ATI), has been observed in many experiments [1]. The angular distributions of the photoelectrons can exhibit pronounced structure [2] which, as we shall show below, reflects the significant alteration of the continuum states in these strong fields. A second remarkable observation in high intensity experiments is the unexpectedly efficient production of *very* high-order harmonic radiation during the laser pulse. For example, harmonics up to the 109th order of a 140 fs Ti-sapphire (806 nm) laser in neon have recently been reported [3].

In this paper we discuss the method we have developed to study the non-perturbative multiphoton processes described above. Recent results will also be presented. The calculations have employed a single-active-electron (SAE) model which explicitly follows the time evolution of one of the valence electrons in the frozen, mean-field of the remaining electrons, the nucleus and the pulsed electromagnetic field. This model has been shown to be quite accurate for the rare gas atoms, which have been the subjects of the majority of experiments to date. The success of the model is at least in part because the neglected double or higher excitations involve states well above the ionization threshold [4]. Most of the processes observed can be understood in terms of the sequential stripping of the atomic electrons. In the next section we will give a description of the SAE model, emphasizing in particular the development of ion-core specific effective potentials.

2 Method and Model

2.1 Time-Dependent Theory of Multiphoton Processes

A many electron atom in an intense, pulsed laser field is a formidable computational problem. This is because the multi-dimensional Hamiltonian

$$H(t) = H_0 + V_I(\{\mathbf{r}_i\}, t), \tag{1}$$

in the time-dependent Schrödinger equation, TDSE

$$i\frac{\partial}{\partial t}\Psi(\{\mathbf{r}_i\}, t) = H(t)\Psi(\{\mathbf{r}_i\}, t) \tag{2}$$

is itself explicitly time dependent (atomic units, $e = m = \hbar = 1$, are used throughout unless otherwise noted). H_0 is the (non-relativistic) atomic Hamiltonian for the electrons with coordinates $\{\mathbf{r}_i\}$, $i = 1, \ldots, n$ and V_I is the interaction between the electrons and the field. For the problems of interest here, the interaction term can be comparable in strength to the Coulombic interactions within the atom. Therefore the method of solution must be able to treat these different interactions on an equal footing. Exact solution of the full multi-electron problem is beyond present computational capabilities, so we have developed a model based on the Hartree-Fock approximation in which the time-dependent wave function is represented by a product of single particle orbitals. This approach provides considerable insight into the excitation dynamics as well as quantitative predictions for the observed multiphoton processes.

The laser is assumed to be linearly polarized along the z-axis and, in the non-perturbative regime, is strong enough to be treated semi-classically, so that

$$V_I(\{\mathbf{r}_i\}, t) = \sum_i z_i \varepsilon_0 f(t) \sin(\omega t). \tag{3}$$

Here ε_0 is the magnitude of the field and ω is the frequency. The pulse envelope function, $f(t)$, is typically chosen to represent either a sine-squared or a trapezoidal pulse. If we are interested in calculating quantities relevant to a specific intensity, $I_0 = c\varepsilon_0^2/8\pi$, we choose a pulse envelope function that rises over several optical cycles to its maximum value (1.0) and is then held constant for 20–30 cycles. The pulse rise must involve at least a few cycles or the calculated results can be contaminated by unphysical transients.

In a linearly polarized field a free electron "quivers" back and forth in response to the oscillating field of the laser with an amplitude given by

$$\alpha_0 = \varepsilon/\omega^2. \tag{4}$$

The amplitude of this motion can greatly exceed the size of the bound electron orbits, which are of the order of $1a_0$. For example, at an intensity of 10^{13} W/cm^2 and a wavelength of 1 μm, the quiver amplitude is almost $10a_0$. An electron freed from an atom in this field will immediately attain this quiver motion on top of its drift motion away from the ion core. The cycle averaged energy of this oscillatory motion is called the ponderomotive energy, U_p, given by

$$U_p = \varepsilon^2/4\omega^2. \tag{5}$$

For the case cited above, the ponderomotive energy is approximately 1 eV. For typical short pulse experiments today, this energy can easily be hundreds of electron volts. Therefore the wave function of a photoelectron in an intense laser field does not resemble that of the normal field-free Coulomb state, but is dressed by the field, becoming, in the absence of a binding potential, a Volkov state [5]. This complex motion of the photoelectrons in the continuum is very difficult to reproduce in terms of the field-free atomic basis functions, so that we have chosen to define our electron wave functions on a finite difference grid. These numerical wave functions have the flexibility to represent both the bound and continuum states in the laser field accurately.

2.2 One-Electron Systems

We have solved the TDSE using both the length and velocity gauges. The equations above are those for the length gauge. Although one gauge is generally found to be more computationally tractable depending on the peak laser intensity and wavelength, both must give the same emission rates when the calculations are converged. We have employed both cylindrical and spherical coordinate systems in our calculations. Again, the choice depends on the laser parameters. By concentrating on linear polarization, we need consider only two spatial dimensions in the description of the electronic state because the azimuthal quantum number (conventionally denoted by m) is conserved. We first present the equations appropriate to a single electron (hydrogenic) system, initially in its ground state, assumed for the moment to be $m = 0$. (See [4] for more details.)

The TDSE for hydrogen in a pulsed laser field using the length gauge is

$$i\frac{\partial}{\partial t}\Psi(\mathbf{r}, t) = \left\{-\frac{1}{2}\nabla^2 - \frac{1}{r} + \varepsilon_0 f(t)\, z \sin(\omega t)\right\}\Psi(\mathbf{r}, t), \tag{6}$$

where the laser field is assumed to be polarized along the z direction. We expand the electron orbital in spherical harmonics

$$\Psi(r, \theta, \phi, t) = \sum_{l=0}^{L} \Phi_l(r, t)\, Y_l^o(\theta, \phi), \tag{7}$$

and discretize the radial coordinate, defining

$$\Phi_l(r, t) \rightarrow \Phi_l(r_j, t) \rightarrow \Phi_l^j(t) \equiv g_l^j(t)/r_j, \tag{8}$$

with $r_j = (j - 0.5)\Delta r$, for $j = 1, \ldots, N_r$. The maximum spherical harmonic, L, the maximum radial grid point, and the grid spacing are all adjusted as required to achieve convergence. In our calculations, Δr is generally of the order of 0.1–0.2a_0, N_r is several thousand and L of the order of 100. Upon substituting the expansion in Eq. (7) into the Lagrange-type functional

$$\mathscr{L} = \left\langle \Psi \left| i\frac{\partial}{\partial t} - H \right| \Psi \right\rangle, \tag{9}$$

using a three point second derivative formula for the radial kinetic energy terms, and requiring that the action associated with this functional be stationary with respect to small variations of the orbital on the grid points, we obtain the following equation for the time evolution:

$$\begin{aligned} i\frac{\partial}{\partial t} g_l^j = & -\frac{1}{2(\Delta r)^2}\left\{c_j g_l^{j+1} + c_{j-1} g_l^{j-1} - 2d_j g_l^j\right\} \\ & + \left\{\frac{l(l+1)}{2r_j^2} - \frac{1}{r_j}\right\} g_l^j + \varepsilon_0 f(t)\, r_j \sin(\omega t)\left\{a_l g_{l+1}^j + a_{l-1} g_{l-1}^j\right\} \end{aligned} \tag{10}$$

$$\equiv (H_0 g)_l^j + (H_I g)_l^j. \tag{11}$$

The coefficients are given by

$$c_j = \frac{j^2}{j^2 - 1/4}, \quad d_j = \frac{j^2 - j + 1/2}{j^2 - j + 1/4}, \quad a_l = \frac{l+1}{\sqrt{(2l+1)(2l+3)}}. \tag{12}$$

The atomic Hamiltonian, H_0, couples radial values j to j and $j \pm 1$ and is diagonal in l while the interaction term H_{I} couples angular momentum l to $l \pm 1$ and is diagonal in j.

The wave function is propagated forward in time via a Peaceman-Rachford alternating-direction, implicit scheme [6] given by

$$g_l^j(t + \Delta t) = [I + iH_o\tau]^{-1}[I + iH_I\tau]^{-1}[I - iH_I\tau][I - iH_o\tau]\, g_l^j(t), \tag{13}$$

with $\tau = \Delta t/2$. The interaction term is evaluated at the midpoint of the integration time step. Generally the time step is of the order of 1/600 of the optical cycle, but can be upto a factor of 30 smaller for calculations with low frequencies and high intensities. The matrices in Eq. (13) are tridiagonal, so all of the matrix multiplications and inversions are accomplished with a small number of vector operations. The propagator is accurate to second order in the time step and the error is proportional to the product of H_0 and H_I. The computational effort is linear in the number of grid points given by the product of L and the number of radial points. On a Cray Y/MP machine (single processor) we can propagate 1.2×10^6 space-time points per cpu second.

2.3 Multi-Electron Systems

2.3.1 Time-Dependent Hartree-Fock

For multi-electron systems, it is not feasible, except possibly in the case of helium, to solve the exact atom-laser problem in $3n$-dimensional space, where n is the number of electrons. One might consider using time-dependent Hartree Fock (TDHF) or the time-dependent local density approximation to represent the state of the system. These approaches lead to at least $n/2$ coupled equations in 3-dimensional space which is much more attractive computationally. For example, in TDHF the wave function for a closed shell system can be approximated by a single Slater determinant of time dependent orbitals,

$$\Psi(\{\mathbf{r}_i\}, t) = A \prod_{i=1}^{n} \varphi_i(\mathbf{r}_i, t), \tag{14}$$

where A is an operator which antisymmetrizes the product wave function. Substituting this expansion into Eq. (9) and minimizing its variation with respect to the time-dependent single particle orbitals we obtain the following set of equations:

$$\begin{aligned} i\frac{\partial}{\partial t}\varphi_i(\mathbf{r}_i, t) = &\left\{ -\frac{1}{2}\nabla_i^2 + \sum_j \int d\mathbf{r}_j \frac{|\varphi_j(\mathbf{r}_j, t)|^2}{|\mathbf{r}_i - r_j|} \right. \\ &\left. - \frac{Z}{r_i} + z_i \varepsilon_0 f(t) \sin(\omega t) \right\} \varphi_i(\mathbf{r}_i, t) \\ &- \sum_j \left\{ \int d\mathbf{r}_j \frac{\varphi_j^*(\mathbf{r}_j, t)\, \varphi_i(\mathbf{r}_j, t)}{|\mathbf{r}_i - \mathbf{r}_j|} \right\} \varphi_j(\mathbf{r}_i, t), \quad i = 1, \ldots, n. \end{aligned} \tag{15}$$

The last term on the right of Eq. (15) is the non-local exchange term. It can be approximated by a local functional of the total density [7] which leads to the time-dependent local density approximation. In either case the equations are nonlinear in the orbitals because the electronic interactions are approximated by density dependent terms. This presents a problem at high intensities when the atom ionizes rapidly [8]. The total electron density near the atom gradually decreases as the electrons are promoted to the continuum, which causes the density-dependent terms in Eq. (15) to diminish. This is qualitatively correct, the instantaneous ionization potential of the system increases as the electrons are ionized, just as in a real atom. However, this change in the ionization potential is continuous, rather than having abrupt jumps as each electron is removed from the atom. This presents a serious problem in the calculation of multiphoton excitation and ionization because the emission rates for all processes of interest depend exponentially on the ionization potential. The problem is most acute if the atom is ionized to an appreciable degree during the rise of the pulse, meaning that the system which experiences the peak intensity is an unphysical, fractionally ionized atom. Calculations on helium using a TDHF wave function have shown that this continuous weakening of the mean field produces very substantial errors in predicted ionization rates [8]. Results for electron and photon spectra would be similarly affected.

2.3.2 Single-Active-Electron Approximation

To avoid this problem, we have developed an approximation to the full multi-electron Schrödinger equation which treats the evolution of each electron individually, but neglects the *dynamic* interelectron interactions [8, 9]. We arrive at this model in the following way. First the single-particle orbitals in the Hartree-Fock wave function are defined in terms of their departure from the initial state,

$$\varphi_i(\mathbf{r}_i, t) \rightarrow \varphi_i^0(\mathbf{r}_i) + \delta_i(\mathbf{r}_i, t), \tag{16}$$

where φ_i^0 is the field-free Hartree-Slater ground state orbital. Putting these orbitals into Eq. (14) and assuming the δ_i are small the leading term in the wave function is

$$\Psi(\{\mathbf{r}_i\}, t) \rightarrow A \sum_j \left\{ [\varphi_j^0(\mathbf{r}_j) + \delta_j(\mathbf{r}_j, t)] \prod_{i \neq j}^{n} \varphi_i^0(\mathbf{r}_i) \right\}. \tag{17}$$

Keeping terms of this order in the TDHF equation Eq. (15) we obtain n uncoupled equations. In each equation only one of the electrons is "active", moving in response to the laser field in the potential generated by the remaining electrons, which are frozen in their initial orbitals. This single-active-electron (SAE) approximation results in a linear Schrödinger equation that is well behaved with respect to obtaining emission rates that are representative of the ionization stage being studied, but suffers from the neglect of pathways involving multiple excitations. In perturbation theory we think of multiphoton excitation in terms of the different virtual transitions, or paths, which lead to the same final state. In particular, we expect a four-photon transition, in addition to various four-photon pathways, will have weak contributions from six- and eight-photon processes. At high intensity, perturbation theory breaks down because the amplitudes of the higher order processes become as large as the lowest order one. The perturbation series no longer converges. However it is still possible, in this regime, to think in terms of the states which become excited during the pulse, but we can no longer ask how many photons were involved in putting a population into a particular state. Our calculations using the SAE model show that, for the rare gas systems and visible or IR wavelengths, the dominant effects seem to be adequately represented by single excitation pathways. The contributions to the overall emission rates from the transitions which represent multi-electron excitation evidently are much less important. Obviously, this will not be the case for all wavelengths and intensities. When multiple excitations are important, for example as they would be in the alkaline earths [10], a more general approach is required.

Calculations using the SAE model are best carried out in spherical coordinates since this allows the use of ℓ-dependent effective potentials. We illustrate our method by presenting calculations for xenon and comparing our results with those from recent experiments. The effective potentials we need are generated by performing Hartree–Slater [7] calculations on the ground and low-lying states of the atom [11]. The p-potential ($l = 1$) comes from the ground state calculation. The $X\alpha$-parameter was varied to reproduce the exact ionization potential. The nodes were removed from the valence p-orbital and the radial Schrödinger equation for this orbital was inverted to give the desired potential [12]. Effective potentials for $l = 0$ and 2 were generated from calculations for the $5p^5 6s$ and $5p^5 5d$ singlet states, respectively. For the excited states, the fine-structure splittings mean that no unique orbital energy is available. Choosing a representative value within the observed energy spread for these configurations provides potentials with eigenenergies reasonably close to the true excited states for all of the singlet configurations $5p^5 ns$ and $5p^5 nd$. For higher angular momenta ($l > 2$) we use the effective potential for $l = 2$. This approach can compensate to some extent for the core relaxation in the excited states for $l \neq 1$ because of the flexibility in choosing the correct orbital energies for these effective potentials. We find this procedure results in quite accurate excitation energies, much better than normally obtained in either Hartree–Fock or Hartree–Slater calculations.

In Fig. 1 we show the xenon effective potentials and their lowest eigenfunctions for $l = 0$ and 1. Both effective potentials become strongly repulsive in the inner core region to mimic the orthogonality constraints on the true orbitals.

In the SAE calculations we solve an equation identical to Eqs. (6) and (10) except that the $1/r$ term is replaced by the proper radial potential corresponding to the particular l-component of the wave function. This does not alter the structure of the coupled equations, and so the propagation is essentially the same as for a hydrogen atom. Note that if we wish to consider the excitation of one of the p-electrons with quantum number m different from zero, the a_l coefficients defined in Eqs. (10–12) become

$$a_l^m = \sqrt{\frac{(\ell - m + 1)(\ell + m + 1)}{(2\ell + 1)(2\ell + 3)}}, \tag{18}$$

and the rest of the calculation is unmodified.

This procedure provides a model of the xenon atom which accounts only for the manifold of singly excited states based on the lowest ionic core, $^2P_{3/2}$. For all rare gases, a second manifold of states converges to the next spin-orbit component of the ion, the $^2P_{1/2}$ state. For example, these two ionization limits in xenon are separated by 1.3 eV corresponding to different total angular momenta, J, of the $5p^5$ configuration. The lower ionization potential is 12.15 eV. We assume that multiphoton excitations into these two manifolds are very weakly coupled so they can be treated separately. This assumption is reasonable because once one of the electrons is excited outside a particular core configuration, transitions

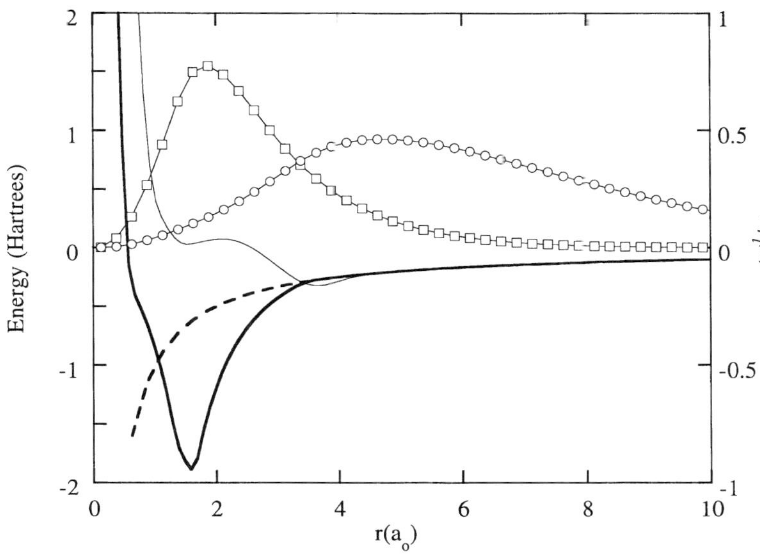

Fig. 1. Effective potentials for $l = 0$ (*light solid line*) and $l = 1$ (*heavy solid line*) for xenon. The $5p$ (*squares*) and $6s$ (*circles*) pseudo-orbital amplitudes are also shown.

into states with the other core requires two electrons to change state. Thus it is within the spirit of the SAE approximation to neglect such excitation pathways. Therefore, we repeat the above procedure to obtain effective potentials for this second set of excited states. Because the emission rates are very sensitive to the magnitude of the ionization potential, this lower limit dominates the dynamics and we can generally ignore the additional manifold of excited states associated with the higher energy core.

We have constructed such core-specific effective potentials for all the rare gases. We have found that for the lighter rare gases, including krypton where the ionic cores are separated by 0.66 eV, the additional manifold of excited states can become important, and, for some wavelengths and intensities, are dominant.

In Fig. 2 we compare the $6s$ orbitals obtained for the two different couplings of the ion core. The difference in the calculations for these two orbitals is that the $X\alpha$-coefficient for the exchange-correlation term in the Hartree–Slater Hamiltonian is varied to shift the calculated orbital energy to agree with the respective binding energy. The Hartree–Slater orbital for the $6s'[^2P_{1/2}$-core] is also shown in Fig. 2. The inner nodes in this orbital are removed to obtain the $6s'$ pseudo-orbital.

Another important effect due to the spin-orbit coupling comes into play whether the upper ionic core is specifically involved or not. This is because the excitation dynamics is very sensitive not only to the ionization potential or binding energy of the active electron but also to m, the projection of the orbital angular momentum along the polarization axis. Since spin-orbit terms are not

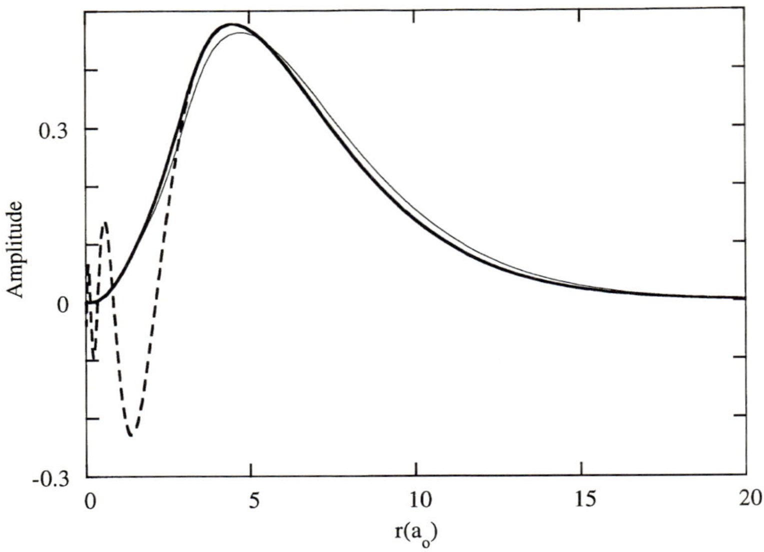

Fig. 2. Orbitals for the $[5p^5(3/2)]6s$ (*light solid line*) and the $[5p^5(1/2)]6s'$ (*heavy solid line*) pseudo-orbitals from which the *s*-wave effective potentials were determined. The $6s'$ Hartree–Slater orbital is also shown (*heavy dashed line*)

included in our Hamiltonian, we solve the Schrödinger equation in *LS*-coupling. This means that for a linearly polarized laser field, m is a good quantum number. Therefore we need to assign weights to the valence orbitals according to their weights in the spin-orbit coupled states. In earlier calculations on xenon, the valence shell p-electrons with $m = 0$ were found to provide the dominant contribution to the ionization rates [9]. There are two $m = 0$ valence electrons in the p-shell, 4/3 associated with the ${}^2P_{3/2}$ core and 2/3 with ${}^2P_{1/2}$. Contributions to the ionization rates and harmonic emission strengths from the other four p-electrons in the valence shell of xenon which have $m = \pm 1$ are found to be unimportant for the intensities considered here. Thus we solve the time-dependent SAE equation for the 4/3 electrons with $m = 0$ that leave the ion in the ${}^2P_{3/2}$ state and the 2/3 electrons with $m = 0$ that leave the ion in the ${}^2P_{1/2}$ state. For the results discussed in this paper we solve Eq. (6) for the orbital, $\varphi^m_{\ell=1}(\mathbf{r}, t)$, with $J = 3/2$ and $m = 0$ in a pulse which rises to its maximum intensity over five optical cycles and then has a constant intensity for the next 20–30 cycles. After the ramp, the transient excitations decay by ionization over the next few cycles. The ionization and photoemission rates are determined during the last part of the pulse. We obtain an ionization rate by monitoring the norm of the wave function. Our finite difference integration of Eq. (6) is carried out in a finite box with absorbing boundaries [13]. As flux reaches the edges of the grid it is removed. The rate at which this occurs is defined to be the ionization rate.

2.4 Calculation of Emission Spectra

2.4.1 Photoemission Spectra

The oscillating laser field distorts the electronic charge density producing a time-dependent dipole in the atom which itself will radiate. The emission strength at a given frequency ω is given by the frequency cubed times the square of the Fourier transform of the total induced dipole [14]. We typically calculate the power spectrum, $\sigma(\omega)$, over the last five cycles of the pulse described above:

$$\sigma(\omega) = \omega^3 \, | d(\omega) |^2 = \omega^3 \left| \frac{1}{T_f - T_i} \sum_j \int_{T_i}^{T_f} dt e^{-i\omega t} \langle \varphi_j(t) | z | \varphi_j(t) \rangle \right|^2 . \tag{19}$$

Contributions to the induced dipole from the different valence orbital calculations are added coherently.

2.4.2 Photoelectron Distributions

If we are interested in the energy and angular distributions of the ejected electrons, we choose a pulse which ramps back down to zero over a few additional cycles (a trapezoidal pulse) and then perform an analysis of the final wave function [15]. We define a window operator

$$W(E_k, n, \gamma) \equiv \frac{\gamma^{2^n}}{(H_0 - E_k)^{2^n} + \gamma^{2^n}} \tag{20}$$

in terms of H_0, the field-free Hamiltonian. The expectation value of W is, to a good approximation, proportional to the total probability of finding the electron with an energy within the interval $E_k \pm \gamma$. The approximation becomes better as n increases, but is quite accurate already for $n = 2$. The energy bin becomes more rectangular as n increases, while its width remains 2γ for any n. As long as γ is small compared to any real structure in the energy distribution, a plot of $\langle W(E_k) \rangle$ vs E_k where $E_{k+1} = E_k + 2\gamma$ will have a shape independent of γ. We have found this to be the most efficient and sensitive method for extracting the energy distributions from the wave function.

This completes our brief description of the computational methods used in these studies. In the following sections some recent results will be presented and discussed. We will cover the calculation of ionization rates, the photoelectron energy distributions, the determination of the residual excited state populations remaining after excitation by a short pulse and finally show some photoemission spectra. The shape of the pulse envelope clearly can affect all these observable quantities. For example, the final state populations are found to be very sensitive to the pulse width and the peak intensity. Such results emphasize the point that in a strong, short pulsed field, the time dependence of the field envelope is reflected in the time evolution of the excitation dynamics. During the pulse, population is moved around among the states of the system and subsequent evolution depends on what has occurred previously. With our time dependent wave function we can explicitly follow the excitation dynamics and understand how the final results are created.

3 Results

3.1 Ionization

We have studied the excitation and ionization of xenon for intensities from 5×10^{12} to above 10^{14} W/cm^2. In Fig. 3 we show the calculated ionization rates for a wavelength of 780 nm. Eight of these 1.6 eV photons are necessary to ionize the atom, leaving the ion in the lower ($J = 3/2$) spin-orbit configuration and nine are needed to reach the $J = 1/2$ threshold. The figure shows that the ionization rate exhibits structure as a function of intensity. This is due to excited states which are ac Stark shifted into resonance, that is to an energy which is an integer number of photons above the initial state. The Stark shifts of highly excited atomic states are generally found to be on the order of the ponderomotive energy of an unbound electron in the laser field which is defined in Eq. (5). At

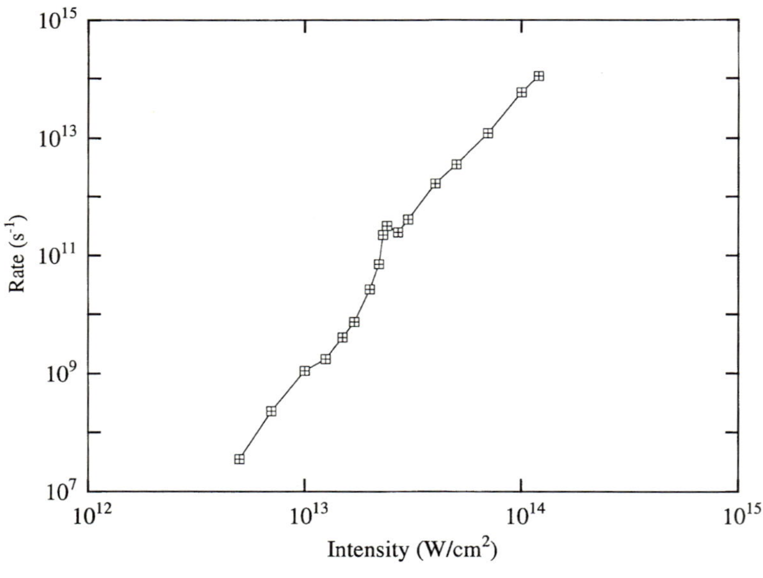

Fig. 3. Calculated SAE ionization rate for xenon at 780 nm

780 nm the shift ranges from a fraction of an electron volt up to almost 7 eV in the intensity range considered. At the upper limit, this shift is many times the photon energy, meaning that, in a pulse which rises to this level, many resonance intensities will be traversed. Thus a state could be in resonance at say seven photons, then again when the intensity has increased enough, at nine photons. The very distinct resonance structure in the ionization rate near 2.5×10^{13} W/cm^2 is due to the eight photon excitation of the Stark shifted $4f$ state. The field-free excitation energy of the $4f$ state is 11.3 eV which is 1.5 eV below eight photons. A ponderomotive shift of 1.5 eV requires an intensity of 2.6×10^{13} W/cm^2, just where the resonance occurs. A second indication that the excited states are playing an important role in the ionization dynamics is that the overall dependence of the rate on intensity is very close to I^5 rather than I^8 as one would expect from perturbation theory for non-resonant ionization. We note that all the xenon excited states are more than five photons above the ground state, so the weaker than expected intensity dependence is not simply attributable to resonances of that order.

In most experiments, lasers have to be focused to achieve the high intensities we are considering here. This means that to compare theory and experiment, we must calculate the atomic ionization rates for a range of intensities, then integrate over the temporal and spatial intensity variation within the focal volume to determine the number of ions produced for a given pulse. In Fig. 4 we show such a comparison for a 140 fs 780 nm pulse. The absolute yield reported

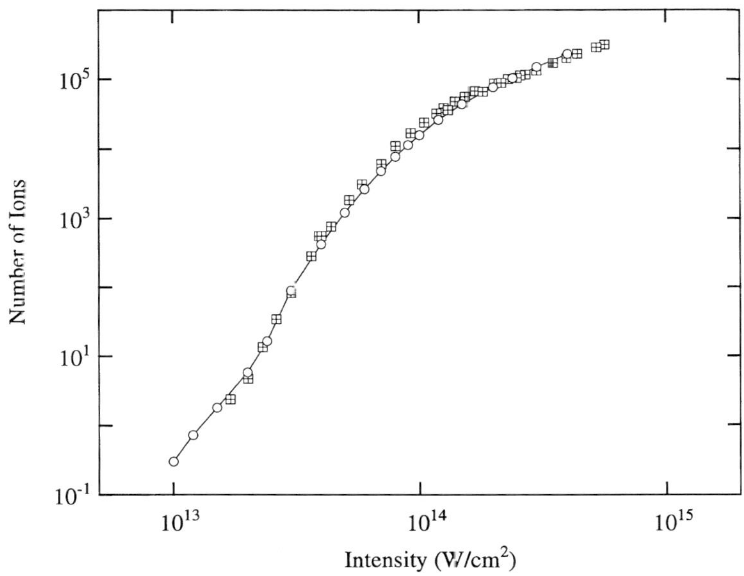

Fig. 4. Calculated (*open circles*) and measured (*crossed boxes*) xenon ion yields for a 780 nm–140 fs laser pulse [16]

in the experiment also involves the ion detection efficiency, so that the comparison is not absolute, but only relative. The agreement is very good over the entire intensity range considered. From Fig. 3 we find that the ionization rate exceeds the inverse of this pulse width around 7×10^{13} W/cm^2 which means that for peak intensities above this value atoms in the central part of the focal volume will be ionized before the peak intensity is reached. Therefore once this intensity is exceeded the ion yield ceases to track the single atom rate but saturates. For a gaussian focus the ion signal rolls over to an $I^{3/2}$ dependence due to the expanding focal volume as the peak intensity is increased above the saturation intensity.

3.2 Population Transfer

The excited states which enhance the ionization of an atom can accumulate a significant population during the pulse which may remain after the field is turned off [17, 18]. This transfer of population from the ground state to the excited states will vary in magnitude depending on the laser wavelength, peak intensity and pulse shape. Here we present calculations of population transfer and resonant ionization in xenon at both 660 and 620 nm. We will concentrate on the $J = 3/2$ channel in what follows. At the longer wavelength, the seven photon channel closes at 2.5×10^{13} W/cm^2. By this we mean that above this intensity the ac Stark shift of the ionization threshold is large enough that it now takes eight photons to reach the continuum rather than the seven required in weaker fields [1]. Pulses with peak intensities higher than this result in "Rydberg trapping", the resonant transfer of population to a broad range of high-lying states. The amount of population transferred depends on both the peak intensity and pulse duration. At 620 nm there are numerous possible six photon resonances to states with p or f angular momenta. We have done a large number of calculations for 40 fs pulses at different peak intensities and have examined the population transferred to these low-lying resonant states as a function of the peak laser intensity.

The pulses that we use, have an intensity envelope given by $\cos^2[\pi^t/2\tau_p]$, τ_p being the full width at half maximum (FWHM). This pulse is very close to a Gaussian pulse of the same FWHM, differing only for intensities far below the peak. It has the advantage of being finite in duration ($-\tau_p < t < \tau_p$), which allows for a precise determination of the final state populations and ionization probabilities. We have compared it to the analogous Gaussian pulses for several cases and have found it to be an excellent approximation.

3.2.1 Rydberg Trapping

We first examine several pulses at a wavelength of 660 nm (1.88 eV). At this wavelength the ponderomotive shift, U_p, of the ionization limit brings the top of the Rydberg series into resonance at $I_{res} = 2.5 \times 10^{13}$ W/cm^2. This seven photon "channel closing" can bring many closely spaced Rydberg states into resonance, each for a short time, during a pulse that has a peak intensity that exceeds I_{res}. At the end of a pulse we project the final state wave function onto bound states of our xenon potential to determine the amount of probability "trapped" in excited states. This population, which is transferred resonantly during the pulse, assumedly does not have sufficient time to ionize out of the excited states. This is quite plausible given the long lifetimes of the high-lying, high ℓ states that can be accessed by seven photons from the $5p$ ground state. We can also think of this, loosely speaking, as the creation of a "Rydberg wave packet" which spends much of the time during the pulse at distances from the nucleus that are too great to allow for efficient ionization [19]. The ionized population is one minus the sum of the ground state population and the trapped population.

Figure 5 shows the ratio of "trapped" to ionized population for several peak intensities and three pulse widths. This ratio is close to zero for peak intensities (I_{peak}) less than I_{res}, as expected. For $I_{peak} > I_{res}$ the ratio grows, and can be as much as one half. The trapped population is found mostly in states with $\ell > 4$ and quantum numbers $n > 7$. Longer pulse widths result in more excitation and ionization, but a lower overall trapped/ionized ratio. Eventually, as the peak intensity is raised and the channel closing occurs earlier in the pulse, the ratio decreases since most of the ionization then occurs non-resonantly. The structure in the curves may be indicative of interference between resonant excitation on the rising and falling edges of the pulse, a subject we will return to in the next section.

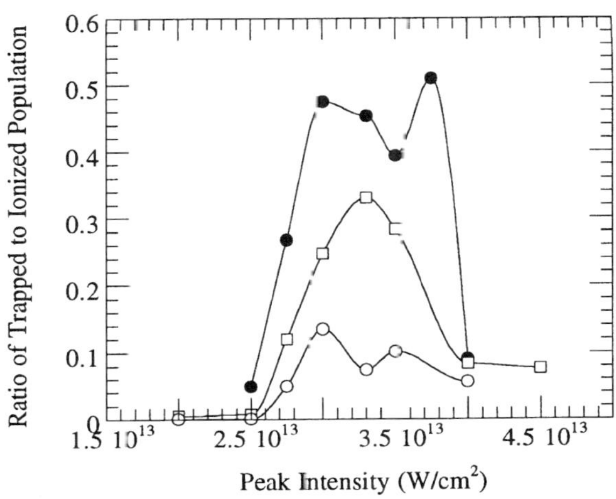

Fig. 5. The trapped/ionized ratio for 50 (*filled circles*), 100 (*open squares*) and 200 (*open circles*) fs FWHM pulses at a wavelength of 620 nm

3.2.2 Single State Trapping

Population can also be transferred resonantly to a single low-lying, relatively isolated excited state during an intense pulse [17a]. We have studied this process for 620 nm (2 eV) photons in detail using 40 fs pulses, a pulse width somewhat shorter than the experiments of de Boer [17a, b] but within current experimental capabilities. A total of 40 pulses with peak intensities between 0.5 and 6.0×10^{13} W/cm^2 were calculated.

At a wavelength of 620 nm, xenon has several six photon resonances in the range between $1\text{–}3 \times 10^{13}$ W/cm^2. These include the $6f$, $5f$ and $4f$ states at $I_{res} = 0.8, 1.1$ and 2.0×10^{13} W/cm^2, respectively, and the $8p$ at 1.4×10^{13} W/cm^2, assuming that these states shift ponderomotively. Other, lower lying states such as the $7p$ and $6s$ (five photons) come into resonance at higher intensities and are not expected to have an ac Stark shift equal to the ponderomotive shift. Figure 6 shows the population remaining in each of these states at the end of the pulse as a function of the peak intensity. Also shown is the total ionization probability. The curves all show a similar shape, with the exception of the 6s which we will comment on below. When $I_{peak} < I_{res}$ little population can be excited to the resonant state. The maximum in the probability to remain in the resonant state occurs at $I_{peak} \sim I_{res}$. And for $I_{peak} > I_{res}$ the probability is smaller and shows oscillations as the peak intensity increases. This is shown in more detail for the $4f$ resonance in Fig. 7.

We can use a simple two state model in the rotating wave approximation (RWA) [20] to explain the main features of Fig. 7. Consider a six photon resonance in a pulse such that $I_{peak} \approx I_{res}$. The ground and resonance states are

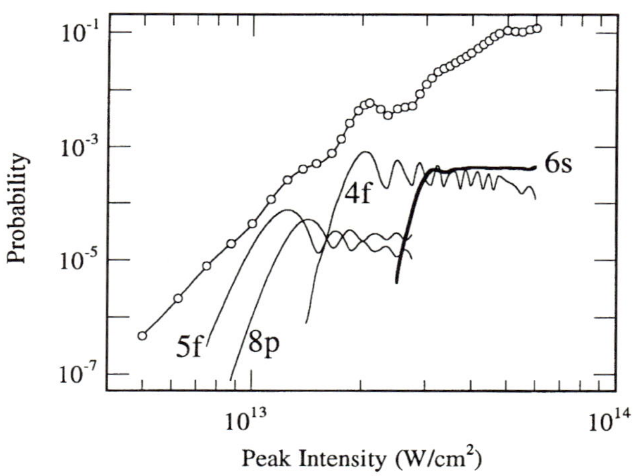

Fig. 6. The population remaining in several bound states at the end of a 40 fs pulse for 620 nm. Also shown is the total amount of ionization (*open circles*)

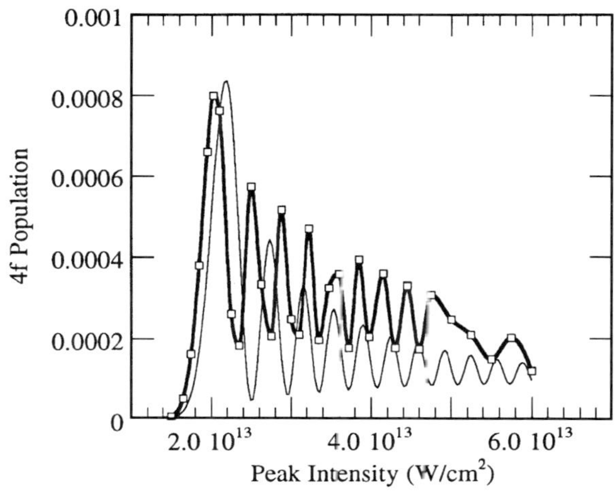

Fig. 7. Population in the 4f state at the end of a 40 fs pulse (the *heavy line* is a spline fit to the calculated points). The lower curve is a two state RWA calculation

connected by an effective six photon Rabi frequency. The detuning from resonance is simply the zero field detuning minus the ac Stark shift, which is proportional to the intensity if the upper state is assumed to shift ponderomotively. For $I_{peak} < I_{res}$ the detuning is always positive and very little population is transferred and hence little remains after the pulse. For $I_{peak} = I_{res}$ the detuning is close to zero for a time proportional to τ_p^2 leading to a relatively large population transfer and, even allowing for ionization, population remains after the pulse. For $I_{peak} > I_{res}$ the detuning goes through zero twice during the pulse for a time proportional to $\sim\tau_p$. If the accumulated phase between the two resonant times is an odd multiple of π, for instance, then destructive interference occurs and the upper state population is zero, in the absence of ionization. That which was promoted to the resonant state during the pulse rise is destroyed when the state comes into resonance the second time. Of course there actually will be some loss from the upper state in the form of one photon ionization. This leads to less than perfect cancellation between the two resonant amplitudes and so the minima in Fig. 7 are not equal to zero. It is interesting to note that we observe oscillations in the upper state population as a function of I_{peak} even though the population in the upper state is always small compared to the lower state. This is because the oscillations come from the interference between the two small amplitudes excited at the two resonant times. Although the two state model gives a qualitative explanation of the oscillations, the RWA result shown in Fig. 7, calculated for a reasonable set of parameters, is deficient in several respects, indicating that the full calculation includes features not captured in the simple two state RWA.

Oscillations in the remaining population are only visible when the lifetime of the upper state is as long or longer than the pulse width. In Fig. 7 this is visible in the damping out of the oscillations at higher peak intensities. In Fig. 6, the 6s state, which comes into resonance at high intensity and which has a short lifetime owing to its low angular momentum, shows no oscillations. This effect can be illustrated by comparing 40 and 100 fs pulses. As shown in Fig. 8, the longer pulse leads to more ionization and *much* less trapping because the lifetime of the 4*f* state is shorter than 100 fs at a few times 10^{13} W/cm^2. The remaining population shows no oscillations for the longer pulse and continues to rise for $I_{peak} > I_{res}$. Again, this behavior is qualitatively reproduced by the two state RWA.

3.3 Above Threshold Ionization

After the laser pulse has passed and some fraction of the wave function has been promoted to the continuum, we want to be able to determine the energy and angular distributions of the emitted electrons. This requires a grid representation of the wave function which is large enough that no flux reaches the boundary while the laser is on. The electron velocities, even at modest intensities, are approximately 1–3 a.u. and the period of the laser in atomic units is of the order of 100. This means that for a few cycle pulse the radial grid must extend to $r > 1000a_0$, and the number of grid points, $N_r > 5000$. These calculations become very time consuming, but turn out to be tractable for intensities and wavelengths such that the ponderomotive energy is below 100 eV. This means we can consider intensities above 10^{14} W/cm^2 even for wavelengths as long as 1 μm.

Having generated the numerical final state wave function using the method presented above, we determine energy distributions for each partial wave using

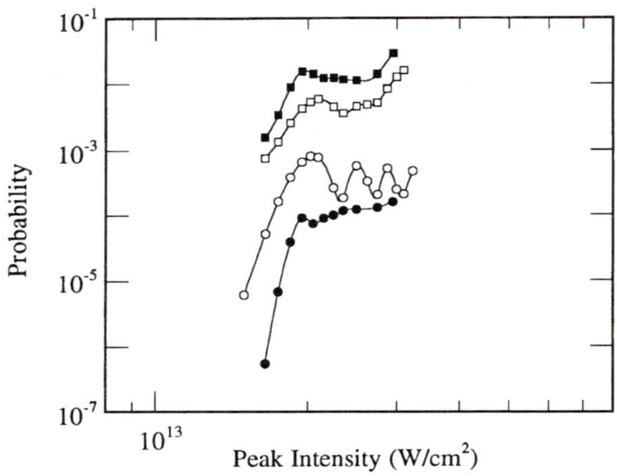

Fig. 8. Ionization (*squares*) and 4f population (*circles*) for 100 (*filled*) and 40 (*open*) fs pulses

the window operator defined in Eq. (20). This provides the energy and angular distributions desired. To illustrate this we show in Fig. 9 the photoelectron energy distribution obtained for a wavelength of 1.053 μm and an intensity of 3×10^{13} W/cm^2. The calculated spectrum shows the feature characteristic of all "long" pulse (> 10 ps) above threshold ionization measurements, a series of peaks separated by the photon energy [1]. The remarkable feature of these data is the very slow decline in intensity as the order of the process increases, which corroborates the non-perturbative nature of the ionization process.

The calculated ATI spectra can be directly compared to measured results as long as the laser intensity is below saturation. Above saturation, spatial averaging becomes increasingly important. In Fig. 10 we show the experimental and theoretical results for two intensities. We find good agreement both in the intensity dependence of the electron yields and in the relative strengths of the peaks for a given intensity. Even the undulations in the peak strengths with energy are reasonably well reproduced. Again, because of the need to account for experimental collection efficiencies, we can only compare the yields on a relative basis. The relative scaling was set for the 18th order peak in the spectrum at the higher experimental intensity.

We have also compared angular distributions of the individual ATI peaks measured and predicted for these same laser conditions [2]. The results showed the appearance of scattering rings on a few of the higher energy peaks. These

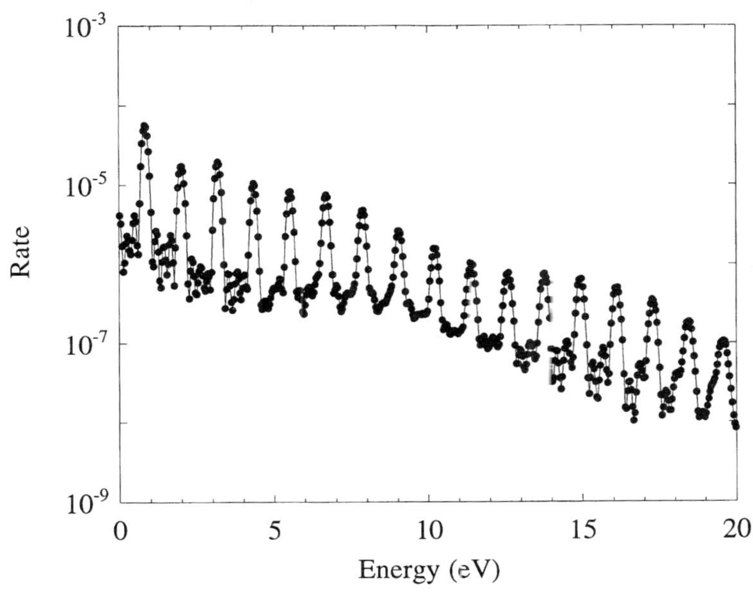

Fig. 9. Calculated photoelectron energy distribution for a laser wavelength of 1.053 μm and an intensity of 3×10^{13} W/cm^2

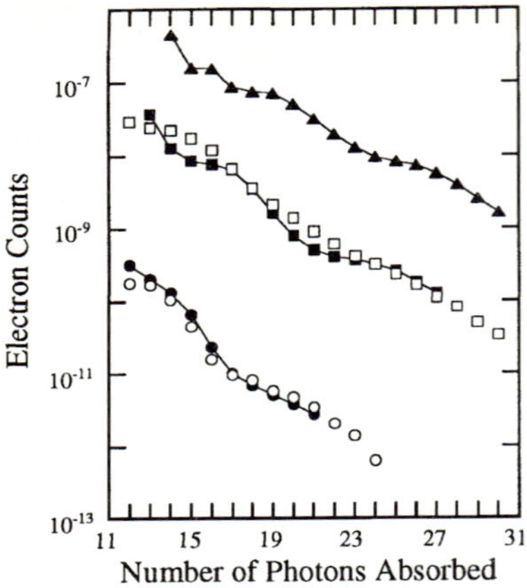

Fig. 10. Photoelectron kinetic energy (ATI) spectrum for Xe at 1.053 μm at 1 (*circles*), 1.5 (*squares*) and 2 (*triangles*) $\times 10^{13}$ W/cm^2: Calculated (*filled symbols*), Measured (*open symbols*) [16]

rings were present for only a narrow range of photoelectron energies, with the window of energies with rings being strongly intensity dependent. The remarkable structures observed can be understood in terms of the dynamics of the electrons within the continuum [2]. This is another signature of the non-perturbative behavior found in this intensity regime.

3.4 Harmonic Generation

The other important emission process discovered in the strong-field, non-perturbative regime, is the production of surprisingly intense and very high order harmonics [3]. Atoms in a laser field will develop a dipole moment which oscillates in phase with the field, much as a free electron would oscillate in such a field. The atomic electrons, however, are confined by their binding potentials which inhibit the free oscillation of the electron, introducing anharmonic components to its motion. Those frequency components which are an integer multiple of the driving field will constructively interfere to produce a polarization field, resulting in the emission of harmonic photons [14, 21]. Because of the inversion symmetry of the atomic system, only the odd harmonics of the incident laser frequency are generated. In the non-perturbative regime, the distortion of the electronic charge distribution is very pronounced, and, as this distribution is driven across the residual ion core potential, the abrupt collision with the short range part of the potential produces anharmonicities in the induced dipole. If the amplitude of the oscillatory motion of the excited charge density is large enough, the encounter with the ion core will be sudden producing frequency components of many orders with comparable strengths. This leads to a photoemission spectrum which has very broad plateau with many equally intense harmonics. The plateau is observed to have a well defined maximum order [22]. The cutoff in the harmonic spectrum is associated with the maximum energy an excited electron can have such that it can be driven by the field to rescatter from the nucleus. If this *energy* is exceeded, the laser field will not be strong enough to prevent the electron from leaving the vicinity of the ion core without rescattering. Since only electrons which are close to the core can experience the strong acceleration by the short range potential, there must be an intensity dependent maximum to the photoemission spectrum [2, 23].

Below we show two harmonic spectra. The first, Fig. 11, was calculated for the same set of parameters used for the ATI calculation shown in Fig. 9. Here we show the strong narrow peaks at the odd harmonic frequencies which are emitted during the last five cycles of a 30-cycle pulse which was ramped over five cycles to an intensity of 3×10^{13} W/cm^2, then held constant thereafter. The spectrum is obtained from the Fourier transform of the time-dependent induced dipole, Eq. (19). The fifth through the 13th harmonics are approximately comparable in strength, with the higher harmonic peaks becoming gradually weaker. The background between the peaks comes from various sources, among them the weak contributions from resonant excitation channels which, over five cycles, experience only a modest amount of destructive interference. The macroscopic emission spectrum from the entire focal volume is dominated by the harmonic components [3, 21].

The second spectrum we show is for a much higher laser intensity where the distribution of harmonic intensities is much more clearly demonstrated. In Fig. 12 we show the spectrum for neon at 806 nm and an intensity of

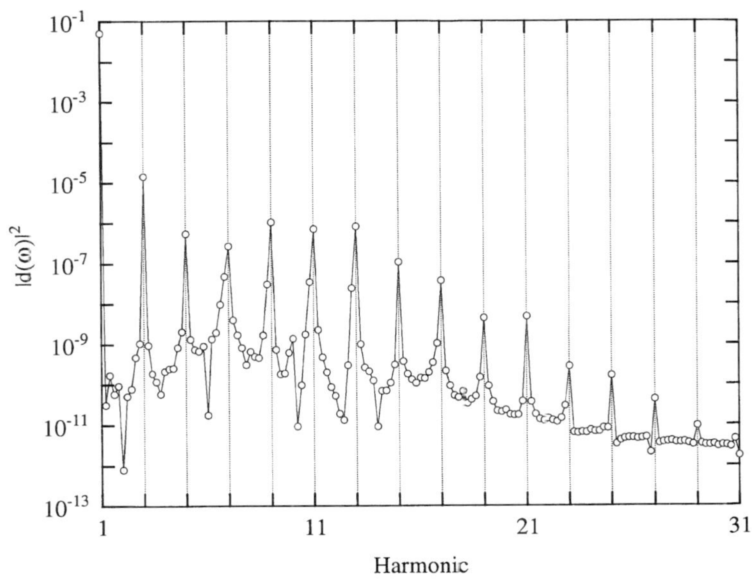

Fig. 11. Xenon harmonic emission strengths for an incident wavelength of 1.053 μm and an intensity of 3×10^{13} W/cm^2

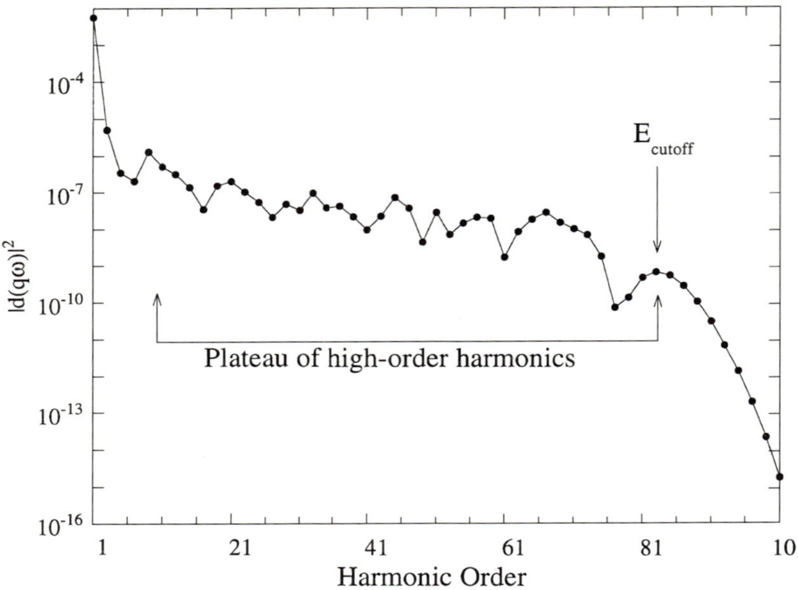

Fig. 12. Neon harmonic emission strengths for a 806 nm laser for an intensity of 6×10^{14} W/cm^2

6×10^{14} W/cm^2. The harmonic plateau and cutoff are noted and the calculated spectrum agrees very well with that reported in [3a]. This shows the emission of photons with energies up to 125 eV, approximately six times the initial binding energy of the outer valence electron. We emphasize the spectrum is that for the neutral atom; only one electron is responsible for this emission. The explanation for the extraordinary range of photon energies is just that, in this laser field, an electron can be accelerated to a velocity large enough to produce the highest harmonics shown, yet not escape from the vicinity of the ion core before rescattering. The strong harmonics produced are coherent and their pulse length is comparable to that of the exciting laser field. This provides a new and promising source of short wavelength radiation.

4 Conclusion

In this paper we have shown the method we have developed to study the response of rare gas atoms to intense, short pulse laser fields. Realizing that the electric field of the laser can become as strong as or stronger than the Coulombic interactions between the atomic electrons, we chose an approach which treats all these forces on an equal footing. This led us to find a way to solve explicitly the time-dependent Schrödinger for the evolution of the electronic wave function throughout the laser pulse. This approach at present is computationally demanding enough that we have restricted our studies to systems where the dominant effects involve the excitation of only one electron at a time. This turns out to be a pretty reasonable assumption in the rare gases as the doubly excited states lie well above the ionization threshold and are expected to play a minor role in the multiphoton processes of interest. We note that if we were to choose wavelengths in the UV or VUV regime, multiple excitations, even in these atoms, could become important. However, most laser experiments have used and continue to use visible and IR frequencies so the single electron picture should work well.

In considering the dynamics of electrons in these strong fields, we also came to the conclusion that standard expansions in terms of the field-free atomic states would converge very slowly, if at all. Therefore we have represented our wave functions on a numerical grid which leads to very efficient, vectorizable computational algorithms. From the calculated wave function we can determine all the observable emission processes for these systems: ion yields, photoelectron energy and angular distributions and photoemission strengths.

The future of this field clearly will rely on extending these, or some other, methods to study two-electron systems and simple molecular systems. There is evidence that electron-electron correlation continues to play a role in excitation dynamics even in very intense fields. The interaction can be small, but it has been observed to yield orders of magnitude enhancements in the production of doubly charged ions for intensities below that at which sequential ionization becomes efficient. In molecules, the transfer of absorbed energy from the electrons to the nuclei which controls the competition between ionization and dissociation is another important and developing field of research.

On the other hand, there is still a considerable amount to learn about atomic processes in more complicated electromagnetic fields, multiple-color fields or pulses with rapidly varying phases and ellipticities. These latter problems can be directly attacked by the methodology described here.

Acknowledgment. The authors would like to acknowledge the contributions to the work reported here by our collaborators in the group of Louis F. DiMauro at Brookhaven National Laboratory, particularly Baorui Yang and Barry Walker, and by Pierre Agostini and Anne L'Huillier from Saclay and Lund, respectively. This work has been carried out under the auspices of the U.S. Department of Energy at the Lawrence Livermore National Laboratory under contract number W-7405-ENG-48.

5 References

1. (a) Agostini P, Fabre F, Mainfray G, Petit G, Rahman NK (1979) Phys Rev Lett 42: 1127; (b) Kruit P, Kimman J, Muller HG, van der Wiel MJ (1983) Phys Rev A 28: 248
2. Yang B, Schafer KJ, Walker B, Kulander KC, Agostini P, DiMauro LF (1993) Phys Rev Lett 71: 3770

3. (a) Macklin J, Kmetec JD, Gordon CL III (1993) Phys Rev Lett 70: 766; (b) L'Huillier A, Balcou P (1993) Phys Rev Lett 70: 774
4. Kulander KC, Schafer KJ, Krause JL (1992) In: Gavrila M (ed) Atoms in intense radiation fields, Academic Press, New York p. 247
5. Volkov DM (1935) Z Phys 94: 250
6. Varga RS (1962) Matrix iterative analysis, Englewood cliffs, N.J., Prentice Hall
7. Herman F, Skillman S (1963) Atomic structure calculations, Englewood Cliffs, NJ, Prentice Hall
8. Kulander KC (1987) Phys Rev A 36: 2726
9. Kulander KC (1988) Phys Rev A 38: 778
10. Zhang J, Lambropoulos P (1995) J Phys B 28: L1
11. Kulander KC, Rescigno TN (1991) Comp Phys Comm 63: 523
12. (a) Kahn L, Baybutt P, Truhlar DG (1976) J Chem Phys 65: 3826; (b) Christiansen PA, Lee YS, Pitzer KS (1979) J Chem Phys 71: 4445
13. (a) Goldberg A, Shore BW (1978) J Phys B 11: 3339; (b) Kulander KC (1987) Phys Rev A 35: 445; (c) Krause JL, Schafer KJ, Kulander KC (1992) Phys Rev A 45: 4998
14. Kulander KC, Shore BW (1989) Phys Rev Lett 62: 524
15. (a) Schafer KJ, Kulander KC (1990) Phys Rev A 42: 5794; (b) Schafer KJ (1991) Comp Phys Comm 63: 427
16. Schafer KJ, Yang B, DiMauro LF, Kulander KC (1993) Phys Rev Lett 70: 1599
17. (a) de Boer MP, Muller HG (1992) Phys Rev Lett 68: 2747; (b) de Boer MP, Noordam LD, Muller HG (1993) Phys Rev A 47: 45; (c) Gibson GN, Freeman RR, McIlrath TJ (1992) Phys Rev Lett 69: 1904; (d) Story JG, Duncan DI, Gallagher TF (1993) Phys Rev Lett 70: 3012; (e) Vrijen RB, Hoogenraad JH, Muller HG, Noordam LD (1993) Phys Rev Lett 70: 3016
18. Schafer KJ, Kulander KC (1994) In Multiphoton Processes (Evans DK, Chin SL, eds.) p. 35, Singapore, World Scientific
19. (a) Noordam LD, ten Wolde A, Muller HG, Lagendijk A, van Linden van den Heuvell HB, (1988) J Phys B 21: L533; (b) Jones RR, Bucksbaum PH (1991) Phys Rev Lett 67: 3215; (c) Stapelfeldt H, Papaioannou DG, Noordam LD, Gallagher TF (1991) Phys Rev Lett 67: 3223
20. Shore BW (1990) The theory of coherent atomic excitation, p. 194, New York, John Wiley & Sons
21. L'Huillier A, Schafer KJ, Kulander KC (1991) J Phys B 24: 3315 and references therein
22. Krause JL, Schafer KJ, Kulander KC (1992) Phys Rev Lett 68: 3535
23. (a) Kulander KC, Schafer KJ, Krause JL (1993) In: Super-Intense Laser-Atom Physics (Piraux B, L'Huillier A, Rzazewski K, eds.) p. 95, New York, Plenum; (b) Corkum PB (1993) Phys Rev Lett 71: 1994

Author Index Volumes 1-86

Aegerter, M. A.: Sol-Gel Chromogenic Materials and Devices. Vol. 85, pp. 149-194.
Ahrland, S.: Factors Contributing to (b)-behavior in Acceptors, Vol 1, pp. 207-220.
Ahrland, S.: Thermodynamics of Complex Formation between Hard and Soft Acceptors and Donors. Vol. 5, pp. 118-149.
Ahrland, S.: Thermodynamics of the Stepwise Formation of Metal-Ion Complexes in Aqueous Solution. Vol. 15, pp. 167-188.
Aisen, P. see Doi, K.: Vol. 70, pp. 1-26.
Alcock, N. W. see Leciejewicz, J.: Vol. 82, pp. 43-84.
Allen, G. C., Warren, K. D.: The Electronic Spectra of the Hexafluoro Complexes of the First Transition Series. Vol. 9, pp. 49-138.
Allen, G. C., Warren, K. D.: The Electronic Spectra of the Hexafluoro Complexes of the Second and Third Transition Series. Vol. 19, pp. 105-165.
Alonso, J. A., Balbas, L. C.: Hardness of Metallic Clusters. Vol. 80, pp. 229-258.
Alonso, J. A., Balbás, L. C.: Simple Density Functional Theory of the Electronegativity and Other Related Properties of Atoms and Ions. Vol. 66, pp. 41-78.
Andersson, L. A., Dawson, J. H.: EXAFS Spectroscopy of Heme-Containing Oxygenases and Peroxidases. Vol. 74, pp. 1-40.
Antanaitis, B. C. see Doi, K.: Vol. 70, pp. 1-26.
Ardon, M., Bino, A.: A New Aspect of Hydrolysis of Metal Ions: The Hydrogen-Oxide Bridging Ligand ($H_3O_2^-$). Vol. 65, pp. 1-28.
Armstrong, F. A.: Probing Metalloproteins by Voltammetry. Vol. 72, pp. 137-221.
Augustynski, J.: Aspects of Photo-Electrochemical and Surface Behavior of Titanium(IV) Oxide. Vol. 69, pp. 1-61.
Averill, B. A.: Fe-S and Mo-Fe-S Clusters as Models for the Active Site of Nitrogenase. Vol. 53, pp. 57-101.

Babel, D.: Structural Chemistry of Octahedral Fluorocomplexes of the Transition Elements. Vol. 3, pp. 1-87.
Bacci, M.: The Role of Vibronic Coupling in the Interpretation of Spectroscopic and Structural Properties of Biomolecules. Vol. 55, pp. 67-99.
Baekelandt, B. G., Mortier, W. J., Schoonheydt, R. A.: The EEM Approach to Chemical Hardness in Molecules and Solids: Fundamentals and Applications. Vol. 80, pp. 187-228.
Baker, E. C., Halstead, G. W., Raymond, K. N.: The Structure and Bonding of 4*f* and 5*f* Series Organometallic Compounds. Vol. 25, pp. 21-66.
Balbás, L. C. see Alonso, J. A.: Vol. 66, pp. 41-78.
Balbás, L. C. see Alonso, J. A.: Vol. 80, pp. 229-258.
Balsenc, L. R.. Sulfur Interaction with Surfaces and Interfaces Studied by Auger Electron Spectrometry. Vol. 39, pp. 83-114.
Banci, L., Bencini, A., Benelli, C., Gatteschi, D., Zanchini, C.: Spectral-Structural Correlations in High-Spin Cobalt(II) Complexes. Vol. 52, pp. 37-86.

Banci, L., Bertini, I., Luchinat, C.: The ^{1}H NMR Parameters of Magnetically Coupled Dimers – The Fe_2S_2 Proteins as an Example. Vol. 72, pp. 113-136.
Baran, E. J. see Müller, A.: Vol. 26, pp. 81-139.
Bartolotti, L. J.: Absolute Electronegativities as Determined from Kohn-Sham Theory. Vol. 66, pp. 27-40.
Bau, R. G. see Teller, R.: Vol. 44, pp. 1-82.
Baughan, E. C.: Structural Radii, Electron-cloud Radii, Ionic Radii and Solvation. Vol. 15, pp. 53-71.
Bayer, E., Schretzmann, P.: Reversible Oxygenierung von Metallkomplexen. Vol. 2, pp. 181-250.
Bearden, A.J., Dunham, W. R.: Iron Electronic Configuration in Proteins: Studies by Mössbauer Spectroscopy. Vol. 8, pp. 1-52.
Bencini, A. see Banci L.: Vol. 52, pp. 37-86.
Benedict, U. see Manes, L.: Vol. 59/60, pp. 75-125.
Benelli, C. see Banci L.: Vol. 52, pp. 37-86.
Benfield, R. E. see Thiel, R. C.: Vol. 81, pp. 1-40.
Bergmann, D., Hinze, J.: Electronegativity and Charge Distribution. Vol. 66, pp. 145-190.
Berners-Price, S.J., Sadler, P. J.: Phosphines and Metal Phosphine Complexes: Relationship of Chemistry to Anticancer and Other Biological Activity. Vol. 70, pp. 27-102.
Bertini, I. see Banci, L.: Vol. 72, pp. 113-136.
Bertini, I., Ciurli, S., Luchinat, C.: The Electronic Structure of FeS Centers in Proteins and Models. A Contribution to the Understanding of Their Electron Transfer Properties. Vol. 83, pp. 1-54.
Bertini, I., Luchinat, C., Scozzafava, A.: Carbonic Anhydrase: An Insight into the Zinc Binding Site and into the Active Cavity Through Metal Substitution. Vol. 48, pp. 45-91.
Bertrand, P.: Application of Electron Transfer Theories to Biological Systems. Vol. 75, pp. 1-48.
Bill, E. see Trautwein, A. X.: Vol. 78, pp. 1-96.
Bino, A. see Ardon, M.: Vol. 65, pp. 1-28.
Blanchard, M. see Linarès, C.: Vol. 33, pp. 179-207.
Blasse, G. see Powell, R. C.: Vol. 42, pp. 43-96.
Blasse, G.: Optical Electron Transfer Between Metal Ions and its Consequences. Vol. 76, pp. 153-188.
Blasse, G.: The Influence of Charge-Transfer and Rydberg States on the Luminescence Properties of Lanthanides and Actinides. Vol. 26, pp. 43-79.
Blasse, G.: The Luminescence of Closed-Shell Transition Metal-Complexes. New Developments. Vol. 42, pp. 1-41.
Blauer, G.: Optical Activity of Conjugated Proteins. Vol. 18, pp. 69-129.
Bleijenberg, K. C.: Luminescence Properties of Uranate Centres in Solids. Vol. 42, pp. 97-128.
Boca, R., Breza, M., Pelikán, P.: Vibronic Interactions in the Stereochemistry of Metal Complexes. Vol. 71, pp. 57-97.
Boeyens, J. C. A.: Molecular Mechanics and the Structure Hypothesis. Vol. 63, pp. 65-101.
Böhm, M. C. see Sen, K. D.: Vol. 66, pp. 99-123.
Bohra, R. see Jain, V. K.: Vol. 52, pp. 147-196.
Bollinger, D. M. see Orchin, M.: Vol. 23, pp. 167-193.
Bominaar, E. L. see Trautwein, A. X.: Vol. 78, pp. 1-96.
Bonnelle, C.: Band and Localized States in Metallic Thorium, Uranium and Plutonium, and in Some Compounds, Studied by X-ray Spectroscopy. Vol. 31, pp. 23-48.
Bose, S. N. see Nag, K.: Vol. 63, pp. 153-197.
Bowler, B. E. see Therien, M. J.: Vol. 75, pp. 109-130.
Bradshaw, A. M., Cederbaum, L. S., Domcke, W.: Ultraviolet Photoelectron Spectroscopy of Gases Adsorbed on Metal Surfaces. Vol. 24, pp. 133-170.
Braterman, P. S.: Spectra and Bonding in Metal Carbonyls. Part A: Bonding. Vol. 10, pp. 57-86.
Braterman, P. S.: Spectra and Bonding in Metal Carbonyls. Part B: Spectra and Their Interpretation. Vol. 26, pp. 1-42.

Bray, R. C., Swann, J. C.: Molybdenum-Containing Enzymes. Vol. 11, pp. 107-144.
Brec, R. see Evain, M.: Vol. 79, pp. 277-306.
Brese, N. E., O'Keeffe, M.: Crystal Chemistry of Inorganic Nitrides. Vol. 79, pp. 307-378.
Breza, M. see Boca, R.: Vol. 71, pp. 57-97.
Briggs, L. R. see Kustin, K.: Vol. 53, pp. 137-158.
Brooks, M. S. S.: The Theory of 5f Bonding in Actinide Solids. Vol. 59/60, pp 263-293.
Brown, D. G. see Wood, J. M.: Vol. 11, pp. 47-105.
Buchanan, B. B.: The Chemistry and Function of Ferredoxin. Vol. 1, pp. 109-148.
Bucher, E. see Campagna, M.: Vol. 30, pp. 99-140.
Buchler, J. W., Dreher, C. Künzel, F. M.: Synthesis and Coordination Chemistry of Noble Metal Porphyrins. Vol. 84, pp. 1-70.
Buchler, J. W., Kokisch, W., Smith, P.D.: Cis, Trans, and *Metal* Effects in Transition Metal Porphyrins. Vol. 34, pp. 79-134.
Bulman, R. A.: Chemistry of Plutonium and the Transuranics in the Biosphere. Vol. 34, pp. 39-77.
Bulman, R. A.: The Chemistry of Chelating Agents in Medical Sciences. Vol. 67, pp. 91-141.
Burdett, J. K.: Some Structural Problems Examined Using the Method of Moments. Vol. 65, pp. 29-90.
Burdett, J. K.: The Shapes of Main-Group Molecules: A Simple Semi-Quantitative Molecular Orbital Approach. Vol. 31, pp. 67-105.

Campagna, M., Wertheim, G. K., Bucher, E.: Spectroscopy of Homogeneous Mixed Valence Rare Earth Compounds. Vol. 30, pp. 99-140.
Carter, R. O. see Müller, A.: Vol. 26, pp. 81-139.
Cauletti, C. see Furlani, C.: Vol. 35, pp. 119-169.
Cederbaum, L. S. see Bradshaw, A. M.: Vol. 24, pp. 133-170.
Cederbaum, L. S. see Schmelcher, P. S.: Vol. 86, pp. 27-62
Ceulemans, A., Vanquickenborne, L. G.: The Epikernel Principle. Vol. 71, pp. 125-159.
Chandrasekhar, V., Thomas, K. R. Justin: Recent Aspects of the Structure and Reactivity of Cyclophosphazenes. Vol. 81, pp. 41-114.
Chandrashekar, T. K. see Ravikanth, M.: Vol. 82, pp. 105-188.
Chang, J. see Therien, M. J.: Vol. 75, pp. 109-130.
Chasteen, N. D.: The Biochemistry of Vanadium. Vol. 53, pp. 103-136.
Chattaraj, P. K. Parr, R. G.: Density Functional Theory of Chemical Hardness. Vol. 80, pp. 11-26..
Cheh, A. M., Neilands, J. P.: The γ-Aminoevulinate Dehydratases Molecular and Environmental Properties. Vol. 29, pp. 123-169.
Chimiak, A., Neilands, J. B.: Lysine Analogues of Siderophores. Vol. 58, pp. 89-96.
Christensen, J. J. see Izatt, R. M.: Vol. 16, pp. 161-189.
Ciampolini, M.: Spectra of 3d Five-Coordinate Complexes. Vol. 6, pp. 52-93.
Ciurli, S. see Bertini, I.: Vol. 83, pp. 1-54.
Clack, D. W., Warren, K. D.: Metal-Ligand Bonding in 3d Sandwich Complexes. Vol. 39, pp. 1-141.
Clarke, M. J., Fackler, P. H.: The Chemistry of Technetium: Toward Improved Diagnostic Agents. Vol. 50, pp. 57-58.
Clarke, M. J., Gaul, J. B.: Chemistry Relevant to the Biological Effects of Nitric Oxide and Metallonitrosyls. Vol. 81, pp. 147-181.
Clarke, R. J. H., Stewart, B.: The Resonance Raman Effect. Review of the Theory and of Applications in Inorganic Chemistry. Vol. 36, pp. 1-80.
Codling, K., Frasinski, L. J. : Molecules in Intense Laser Fields: an Experimental Viewpoint. Vol. 86, pp. 1-26.
Cohen, I. A.: Metal-Metal Interactions in Metalloporphyrins, Metalloproteins and Metalloenzymes. Vol. 40, pp. 1-37.
Connett, P. H., Wetterhahn, K. E.: Metabolism of the Carcinogen Chromate by Cellular Constituents.

Vol. 54, pp. 93-124.
Cook, D. B.: The Approximate Calculation of Molecular Electronic Structures as a Theory of Valence. Vol. 35, pp. 37-86.
Cooper, S. R., Rawle, S. C.: Crown Thioether Chemistry. Vol. 72, pp. 1-72.
Cotton, F. A., Walton, R. A.: Metal-Metal Multiple Bonds in Dinuclear Clusters. Vol. 62, pp. 1-49.
Cox, P. A.: Fractional Parentage Methods for Ionisation of Open Shells of *d* and *f* Electrons. Vol. 24, pp. 59-81.
Cras, J. A. see Willemse, J.: Vol. 28, pp. 83-126.
Cremer, D. see Frenking, G.: Vol. 73, pp. 17-96.
Crichton, R. R.: Ferritin. Vol. 17, pp. 67-134.

Dance, J.-M. see Tressaud, A.: Vol. 52, pp. 87-146.
Darriet, J. see Drillon, M.: Vol. 79, pp. 55-100.
Daul, C., Schläpfer, C. W., von Zelewsky, A.: The Electronic Structure of Cobalt(II) Complexes with Schiff Bases and Related Ligands. Vol. 36, pp. 129-171.
Dawson, J. H. see Andersson, L. A.: Vol. 74, pp. 1-40.
Deeth, R. J.: Computational Modelling of Transition Metal Centres. Vol. 82, pp. 1-42.
Degen, J. see Schmidtke, H.-H.: Vol. 71, pp. 99-124.
Dehnicke, K., Shihada, A.-F.: Structural and Bonding Aspects in Phosphorus Chemistry-Inorganic Derivates of Oxohalogeno Phosphoric Acids. Vol. 28, pp. 51-82.
Denning, R. G.: Electronic Structure and Bonding in Actinyl Ions. Vol. 79, pp. 215-276.
Diehn, B. see Doughty, M. J.: Vol. 41, pp. 45-70.
Diemann, E. see Müller, A.: Vol. 14, pp. 23-47.
Dirken, M. W. see Thiel, R. C.: Vol. 81, pp. 1-40.
Dobiáš, B.: Surfactant Adsorption on Minerals Related to Flotation. Vol. 56, pp. 91-147.
Doi, K., Antanaitis, B. C., Aisen, P.: The Binuclear Iron Centers of Uteroferrin and the Purple Acid Phosphatases. Vol. 70, pp. 1-26.
Domcke, W. see Bradshaw, A. M.: Vol. 24, pp. 133-170.
Dophin, D. see Morgan, B.: Vol. 64, pp. 115-204.
Doughty, M. J., Diehn, B.: Flavins as Photoreceptor Pigments for Behavioral Responses. Vol. 41, pp. 45-70.
Drago, R. S.: Quantitative Evaluation and Prediction of Donor-Acceptor Interactions. Vol. 15, pp. 73-139.
Dreher, C. see Buchler, J. W.: Vol. 84, pp. 1-70.
Drillon, M., Darriet, J.: Progress in Polymetallic Exchange-Coupled Systems, some Examples in Inorganic Chemistry. Vol. 79, pp. 55-100.
Dubhghaill, O. M. Ni, Sadler, P. J.: The Structure and Reactivity of Arsenic Compounds. Biological Activity and Drug Design. Vol. 78, pp. 129-190.
Duffy, J. A.: Optical Electronegativity and Nephelauxetic Effect in Oxide Systems. Vol. 32, pp. 147-166.
Dunham, W. R. see Bearden, A. J.: Vol. 8, pp. 1-52.
Dunn, M. F.: Mechanisms of Zinc Ion Catalysis in Small Molecules and Enzymes. Vol. 23, pp. 61-122.

Eatough, D. J. see Izatt, R. M.: Vol. 16, pp. 161-189.
Eller, P. G. see Ryan, R. R.: Vol. 46, pp. 47-100.
Emmerling, A. see Fricke, J.: Vol. 77, pp. 37-88.
Emsley, E.: The Composition, Structure and Hydrogen Bonding of the ß-Diketones. Vol. 57, pp. 147-191.
Englman, R.: Vibrations in Interaction with Impurities. Vol. 43, pp. 113-158.

Epstein, I. R., Kustin, K.: Design of Inorganic Chemical Oscillators. Vol. 56, pp. 1-33.
Ermer, O.: Calculations of Molecular Properties Using Force Fields. Applications in Organic Chemistry. Vol. 27, pp. 161-211.
Ernst, R. D.: Structure and Bonding in Metal-Pentadienyl and Related Compounds. Vol. 57, pp. 1-53.
Erskine, R. W., Field, B. O.: Reversible Oxygenation. Vol. 28, pp. 1-50.
Evain, M., Brec, R.: A New Approach to Structural Description of Complex Polyhedra Containing Polychalcogenide Anions. Vol. 79, pp. 277-306.
Fackler, P. H. see Clarke M. J.: Vol. 50, pp. 57-58.???

Fajans, K.: Degrees of Polarity and Mutual Polarization of Ions in the Molecules of Alkali Fluorides, SrO and BaO. Vol. 3, pp. 88-105.
Fee, J. A.: Copper Proteins - Systems Containing the „Blue" Copper Center. Vol. 23, pp. 1-60.
Feeney, R. E., Komatsu, S. K.: The Transferrins. Vol. 1, pp. 149-205.
Felsche, J.: The Crystal Chemistry of the Rare-Earth Silicates. Vol. 13, pp. 99-197.
Ferreira, R.: Paradoxical Violations of Koopmans' Theorem, with Special Reference to the 3d Transition Elements and the Lanthanides. Vol. 31, pp. 1-21.
Fichtinger-Schepman, A. M. J. see Reedijk, J.: Vol. 67, pp. 53-89.
Fidelis, I. K., Mioduski, T.: Double-Double Effect in the Inner Transition Elements. Vol. 47, pp. 27-51.
Field, B. O. see Erskine, R. W.: Vol. 28, pp. 1-50.
Fischer, J. see Mathey, F.: Vol. 55, pp. 153-201
Follmann, H. see Lammers, M.: Vol. 54, pp. 27-91.
Fournier, J. M., Manes, L.: Actinide Solids. 5f Dependence of Physical Properties. Vol. 59/60, pp. 1-56.
Fournier, J. M.: Magnetic Properties of Actinide Solids. Vol. 59/60, pp. 127-196.
Fraga, S., Valdemoro, C.: Quantum Chemical Studies on the Submolecular Structure of the Nucleic Acids. Vol. 4, pp. 1-62.
Frasinski, L. J.: see Codling, K., Vol. 85, pp. 1-26
Fraústo da Silva, J. J. R., Williams, R. J. P.: The Uptake of Elements by Biological Systems. Vol. 29, pp. 67-121.
Frenking, G. see Jørgensen, C. K.: Vol. 73, pp. 1-16.
Frenking, G., Cremer, D.: The Chemistry of the Noble Gas Elements Helium, Neon, and Argon – Experimental Facts and Theoretical Predictions. Vol. 73, pp. 17-96.
Fricke, B.: Superheavy Elements. Vol. 21, pp. 89-144.
Fricke, J., Emmerling, A.: Aerogels-Preparation, Properties, Applications. Vol. 77, pp. 37-88.
Friebel, C. see Reinen, D.: Vol. 37, pp. 1-60.
Friedrich, H.: Field Induced Chaos and Chaotic Scattering. Vol. 86, pp. 97-124
Friesen, C. see Keppler, B. K.: Vol. 78, pp. 97-128.
Fuhrhop, J.-H.: The Oxidation States and Reversible Redox Reactions of Metalloporphyrins. Vol. 18, pp. 1-67.
Furlani, C., Cauletti, C.: He(I) Photoelectron Spectra of *d*-metal Compounds. Vol. 35, pp. 119-169.

Gallagher, T. F. : Microwave Multiphoton Exitation and Ionization. Vol. 86, pp. 125-148
Galland, P. see Russo, V. E. A.: Vol. 41, pp. 71-110.
Galván, M. see Gázquez, J. L.: Vol. 66, pp. 79-98.
Gatteschi, D. see Banci L.: Vol. 52, pp. 37-86.
Gaul, J. B. see Clarke, M. J.: Vol. 81, pp. 147-181.
Gavezzotti, A. see Simonetta, M.: Vol. 27, pp. 1-43.
Gázquez, J. L., Vela, A., Galván, M.: Fukui Function, Electronegativity and Hardness in the Kohn-Sham Theory. Vol. 66, pp. 79-98.
Gazquéz, J. L.: Hardness and Softness in Densitiy Functional Theory. Vol 80, pp. 27-44.

Gerloch, M., Harding, J. H., Woolley, R. G.: The Context and Application of Ligand Field Theory. Vol. 46, pp. 1-46.
Ghijsen, J. see Naegele, J. R.: Vol. 59/60, pp. 197-262.
Gilbert, T. R. see Kustin, K.: Vol. 53, pp. 137-158.
Gillard, R. D., Mitchell, P. R.: The Absolute Configuration of Transition Metal Complexes. Vol. 7, pp. 46-86.
Gleitzer, C., Goodenough, J. B.: Mixed-Valence Iron Oxides. Vol. 61, pp. 1-76.
Gliemann, G., Yersin, H. Spectroscopic Properties of the Quasi One-Dimensional Tetracyanoplatinate(II) Compounds. Vol. 62, pp. 87-153.
Golovina, A. P., Zorov, N. B., Runov, V. K.: Chemical Luminescence Analysis of Inorganic Substances. Vol. 47, pp. 53-119.
Goodenough, J. B. see Gleitzer, C.: Vol. 61, pp. 1-76.
Grätzel, M. see Kiwi, J.: Vol. 49, pp. 37-125.
Gray, H. B. see Therien, M. J.: Vol. 75, pp. 109-130.
Green, J. C.: Gas Phase Photoelectron Spectra of *d*- and *f*-Block Organometallic Compounds. Vol. 43, pp. 37-112.
Grenier, J. C., Pouchard, M., Hagenmuller, P.: Vacancy Ordering in Oxygen-Deficient Perovskite-Related Ferrites. Vol. 47, pp. 1-25.
Grice, M. E. see Politzer, P.: Vol. 80, pp. 101-114.
Griffith, J. S.: On the General Theory of Magnetic Susceptibilities of Polynuclear Transitionmetal Compounds. Vol. 10, pp. 87-126.
Grisham, C. M. see Mildvan, A. S.: Vol. 20, pp. 1-21.
Gubelmann, M.H., Williams, A. F.: The Structure and Reactivity of Dioxygen Complexes of the Transition Metals. Vol. 55, pp. 1-65.
Güdel, H. U. see Ludi, A.: Vol. 14, pp. 1-21.
Guilard, R., Lecomte, C., Kadish, K. M.: Synthesis, Electrochemistry, and Structural Properties of Porphyrins with Metal-Carbon Single Bonds and Metal-Metal Bonds. Vol. 64, pp. 205-268.
Guillaumont, R. see Hubert, S.: Vol. 34, pp. 1-18.
Gütlich, P.: Spin Crossover in Iron(II)-Complexes. Vol. 44, pp. 83-195.
Gutmann, V. see Mayer, U.: Vol. 12, pp. 113-140.
Gutmann, V., Mayer, H.: Application of the Functional Approach to Bond Variations Under Pressure. Vol. 31, pp. 49-66.
Gutmann, V., Mayer, U.: Redox Properties: Changes Effected by Coordination. Vol. 15, pp. 141-166.
Gutmann, V., Mayer, U.: Thermochemistry of the Chemical Bond. Vol. 10, pp. 127-151.

Häder, D.-P. see Nultsch, W.: Vol. 41, pp. 111-139.
Hagenmuller, P. see Grenier, J. C.: Vol. 47, pp. 1-25.
Hale, J. D. see Williams, R. J. P.: Vol. 1, pp. 249-281.
Hale, J. D. see Williams, R. J. P.: Vol. 15, pp. 1 and 2.
Hall, D. I., Ling, J. H., Nyholm, R. S.: Metal Complexes of Chelating Olefin-Group V Ligands. Vol. 15, pp. 3-51.
Halstead, G. W. see Baker, E. C.: Vol. 25, pp. 21-66.
Hanack, M. see Schultz, H.: Vol. 74, pp. 41-146.
Harding, J. H. see Gerloch, M.: Vol. 46, pp. 1-46.
Harnung, S. E., Schäffer, C. E.: Phase-fixed 3-Γ Symbols and Coupling Coefficients for the Point Groups. Vol. 12, pp. 201-255.
Harnung, S. E., Schäffer, C. E.: Real Irreducible Tensorial Sets and their Application to the Ligand-Field Theory. Vol. 12, pp. 257-295.
Hathaway, B. J.: A New Look at the Stereochemistry and Electronic Properties of Complexes of the Coppes (II) Ion. Vol. 57, pp. 55-118.
Hathaway, B. J.: The Evidence for „Out-of-the Plane" Bonding in Axial Complexes of the Copper(II)

Ion. Vol. 14, pp. 49-67.

Hawes, J. C. see Mingos, D. M. P.: Vol. 63, pp. 1-63.

Hellner, E. E.: The Frameworks (Bauverbände) of the Cubic Structure Types. Vol. 37, pp. 61-140.

Hemmerich, P., Michel, H., Schug, C., Massey, V.: Scope and Limitation of Single Electron Transfer in Biology. Vol. 48, pp. 93-124.

Henry, M., J. P. Jolivet, Livage, J.: Aqueous Chemistry of Metal Cations: Hydrolysis, Condensation and Complexation. Vol. 77, pp. 153-206.

Hider, R. C.: Siderophores Mediated Absorption of Iron. Vol. 57, pp. 25-88.

Hill, H. A. O., Röder, A., Williams, R. J. P.: The Chemical Nature and Reactivity of Cytochrome P-450. Vol. 8, pp. 123-151.

Hilpert, K.: Chemistry of Inorganic Vapors. Vol. 73, pp. 97-198.

Hinze, J. see Bergmann, D.: Vol. 66, pp. 145-190.

Hoffman, B. M., Natan, M. J., Nocek, J. M., Wallin, S. A.: Long-Range Electron Transfer Within Metal-Substituted Protein Complexes. Vol. 75, pp. 85-108.

Hoffmann, B. M. see Ibers, J. A.: Vol. 50, pp. 1-55.

Hoffmann, D. K., Ruedenberg, K., Verkade, J. G.: Molecular Orbital Bonding Concepts in Polyatomic Molecules – A Novel Pictorial Approach. Vol. 33, pp. 57-96.

Hogenkamp, H. P. C., Sando, G. N.: The Enzymatic Reduction of Ribonucleotides. Vol. 20, pp. 23-58.

Hubert, S., Hussonois, M., Guillaumont, R.: Measurement of Complexing Constants by Radiochemical Methods. Vol. 34, pp. 1-18.

Hudson, R. F.: Displacement Reactions and the Concept of Soft and Hard Acids and Bases. Vol. 1, pp. 221-223.

Hulliger, F.: Crystal Chemistry of Chalcogenides and Pnictides of the Transition Elements. Vol. 4, pp. 83-229.

Hussonois, M. see Hubert, S.: Vol. 34, pp. 1-18.

Hyde, B. G. see Makovicky, E.: Vol. 46, pp. 101-170.

Hyde, B. G. see O'Keeffe, M.: Vol. 61, pp. 77-144.

Ibers, J. A., Pace, L. J., Martinsen, J., Hoffmann, B. M.: Stacked Metal Complexes: Structures and Properties. Vol. 50, pp. 1-55.

Ingraham, L. L. see Maggiora, G. M.: Vol. 2, pp. 126-159.

Iqbal, Z.: Intra- and Inter-Molecular Bonding and Structure of Inorganic Pseudohalides with Triatomic Groupings. Vol. 10, pp. 25-55.

Izatt, R. M., Eatough, D. J., Christensen, J. J.: Thermodynamics of Cation-Macrocyclic Compound Interaction. Vol. 16, pp. 161-189.

Jain, V. K., Bohra, R., Mehrotra, R. C.: Structure and Bonding in Organic Derivatives of Antimony(V). Vol. 52, pp. 147-196.

Jerome-Lerutte, S.: Vibrational Spectra and Structural Properties of Complex Tetracyanides of Platinum, Palladium and Nickel. Vol. 10, pp. 153-166.

Johnston, R. L. see Mingos, D. M. P.: Vol. 68, pp. 29-87.

Jolivet, J. P. see Henry, M.: Vol. 77, pp. 153-206.

Jørgensen, C. K. see Müller, A.: Vol. 14, pp. 23-47.

Jørgensen, C. K. see Reisfeld, R.: Vol. 49, pp. 1-36.

Jørgensen, C. K. see Reisfeld, R.: Vol. 69, pp. 63-96.

Jørgensen, C. K. see Reisfeld, R.: Vol. 77, pp. 207-256.

Jørgensen, C. K., Frenking, G.: Historical, Spectroscopic and Chemical Comparison of Noble Gases. Vol. 73, pp. 1-16.

Jørgensen, C. K., Kauffmann, G. B.: Crookes and Marignac - A Centennial of an Intuitive and Pragmatic Appraisal of „Chemical Elements" and the Present Astrophysical Status of

Nucleosynthesis and „Dark Matter". Vol. 73, pp. 227-254.
Jørgensen, C. K., Reisfeld, R.: Uranyl Photophysics. Vol. 50, pp. 121-171.
Jørgensen, C. K.: Deep-Lying Valnce Orbitals and Problems of Degeneracy and Intensitites in Photo-Electron Spectra. Vol. 30, pp. 141-192.
Jørgensen, C. K.: Electric Polarizability, Innocent Ligands and Spectroscopic Oxidation States. Vol. 1, pp. 234-248.
Jørgensen, C. K.: Heavy Elements Synthesized in Supernovae and Detected in Peculiar A-type Stars. Vol. 73, pp. 199-226.
Jørgensen, C. K.: Luminescence of Cerium (III) Inter-Shell Transitions and Scintillator Action. Vol: 85, pp. 195-214.
Jørgensen, C. K.: Narrow Band Thermoluminescence (Candoluminescence) of Rare Earths in Auer Mantles. Vol. 25, pp. 1-20.
Jørgensen, C. K.: Partly Filled Shells Constituting Anti-bonding Orbitals with Higher Ionization Energy than Their Bonding Counterparts. Vol. 22, pp. 49-81.
Jørgensen, C. K.: Photo-Electron Spectra of Non-Metallic Solids and Consequences for Quantum Chemistry. Vol. 24, pp. 1-58.
Jørgensen, C. K.: Predictable Quarkonium Chemistry. Vol. 34, pp. 19-38.
Jørgensen, C. K.: Recent Progress in Ligand Field Theory. Vol. 1, pp. 3-31.
Jørgensen, C. K.: Relationship Between Softness, Covalent Bonding, Ionicity and Electric Polarizability. Vol. 3, pp. 106-115.
Jørgensen, C. K.: The Conditions for Total Symmetry Stabilizing Molecules, Atoms, Nuclei and Hadrons. Vol. 43, pp. 1-36.
Jørgensen, C. K.: The Inner Mechanism of Rare Earths Elucidated by Photo-Electron Spectra. Vol. 13, pp. 199-253.
Jørgensen, C. K.: Valence-Shell Expansion Studied by Ultra-violet Spectroscopy. Vol. 6, pp. 94-115.
Justin, K. R. see Chandrasekhar, V.: Vol. 81, pp. 41-114.

Kadish, K. M. see Guilard, R.: Vol. 64, pp. 205-268.
Kahn, O.: Magnetism of the Heteropolymetallic Systems. Vol. 68, pp. 89-167.
Kalyanasundaram, K. see Kiwi, J.: Vol. 49, pp. 37-125.
Kauffmann, G. B. see Jørgensen, C. K.: Vol. 73, pp. 227-254.
Keijzers, C. P. see Willemse, J.: Vol. 28, pp. 83-126.
Kemp, T. J. see Leciejewicz, J.: Vol. 82, pp. 43-84.
Keppler, B. K., Friesen, C., Moritz, H. G., Vongerichten, H., Vogel, E.: Tumor-Inhibiting Bis (ß-Diketonato) Metal Complexes. Budotitane, cis-Diethoxybis (1-phenylbutane-1,3-dionato) titanium (IV). Vol. 78, pp. 97-128.
Kimura, T.: Biochemical Aspects of Iron Sulfur Linkage in None-Heme Iron Protein, with Special Reference to „Adrenodoxin". Vol. 5, pp. 1-40.
Kitagawa, T., Ozaki, Y.: Infrared and Raman Spectra of Metalloporphyrins. Vol. 64, pp. 71-114.
Kiwi, J., Kalyanasundaram, K., Grätzel, M.: Visible Light Induced Cleavage of Water into Hydrogen and Oxygen in Colloidal and Microheterogeneous Systems. Vol. 49, pp. 37-125.
Kjekshus, A., Rakke, T.: Considerations on the Valence Concept. Vol. 19, pp. 45-83.
Kjekshus, A., Rakke, T.: Geometrical Considerations on the Marcasite Type Structure. Vol. 19, pp. 85-104.
Kokisch, W. see Buchler, J. W.: Vol. 34, pp. 79-134.
Komatsu, S. K. see Feeney, R. E.: Vol. 1, pp. 149-206.
Komorowski, L.: Hardness Indices for Free and Bonded Atoms. Vol. 80, pp. 45-70.
König, E.: Nature and Dynamics of the Spin-State Interconversions in Metal Complexes. Vol. 76, pp. 51-152.
König, E.: The Nephelauxelic Effect. Calculation and Accuracy of the Interelectronic Repulsion Parameters I. Cubic High-Spin d_2, d_3, d_7 and d_8 Systems. Vol. 9, pp. 175-212.

Köpf, H. see Köpf-Maier, P.: Vol. 70, pp. 103-185.

Köpf-Maier, P., Köpf, H.: Transition and Main-Group Metal Cyclopentadienyl Complexes: Preclinical Studies on a Series of Antitumor Agents of Different Structural Type. Vol. 70, pp. 103-185.

Koppikar, D. K., Sivapullaiah, P. V., Ramakrishnan, L., Soundararajan, S.: Complexes of the Lanthanides with Neutral Oxygen Donor Ligands. Vol. 34, pp. 135-213.

Kóren, B. see Valach, F.: Vol. 55, pp. 101-151.

Krause, R.: Synthesis of Ruthenium(II) Complexes of Aromatic Chelating Heterocycles: Towards the Design of Luminescent Compounds. Vol. 67, pp. 1-52.

Krumholz, P.: Iron(II) Diimine and Related Complexes. Vol. 9, pp. 139-174.

Kubas, G. J. see Ryan, R. R.: Vol. 46, pp. 47-100.

Kuki, A.: Electronic Tunneling Paths in Proteins. Vol. 75, pp. 49-84.

Kulander, K. C., Schafer, K. J. : Time-Dependent Calculations of Electron and Photon Emission from an Atom in an Intense Laser Field. Vol. 86, pp. 149-172

Künzel, F. M. see Buchler, J. W.: Vol. 84, pp. 1-70.

Kustin, K. see Epstein, I. R.: Vol. 56, pp. 1-33.

Kustin, K., McLeod, G. C., Gilbert, T. R., Briggs, LeB. R., 4th.: Vanadium and Other Metal Ions in the Physiological Ecology of Marine Organisms. Vol. 53, pp. 137-158.

Labarre, J. F.: Conformational Analysis in Inorganic Chemistry: Semi-Empirical Quantum Calculation vs. Experiment. Vol. 35, pp. 1-35.

Lammers, M., Follmann, H.: The Ribonucleotide Reductases: A Unique Group of Metalloenzymes Essential for Cell Proliferation. Vol. 54, pp. 27-91.

Leciejewicz, J., Alcock, N. W., Kemp, T. J.: Carboxylato Complexes of the Uranyl Ion: Effects of Ligand Size and Coordination Geometry Upon Molecular and Crystal Structure. Vol. 82, pp. 43-84.

Lecomte, C. see Guilard, R.: Vol. 64, pp. 205-268.

Lee, Y. J. see Scheidt, W. R.: Vol. 64, pp. 1-70.

Lehmann, H. see Schultz, H.: Vol. 74, pp. 41-146.

Lehn, J.-M.: Design of Organic Complexing Agents. Strategies Towards Properties. Vol. 16, pp. 1-69.

Licoccia, S. Paolesse, R.: Metal Complexes of Corroles and Other Corrinoids. Vol. 84, pp. 71-134

Linarès, C., Louat, A., Blanchard, M.: Rare-Earth Oxygen Bonding in the $LnMO_4$ Xenotime Structure. Vol. 33, pp. 179-207.

Lindskog, S.: Cobalt(II) in Metalloenzymes. A Reporter of Structure-Function Relations. Vol. 8, pp. 153-196.

Ling, J. H. see Hall, D. I.: Vol. 15, pp. 3-51.

Liu, A., Neilands, J. B.: Mutational Analysis of Rhodotorulic Acid Synthesis in *Rhodotorula philimanae.* Vol. 58, pp. 97-106.

Livage, J. see Henry, M.: Vol. 77, pp. 153-206.

Livorness, J., Smith, T.: The Role of Manganese in Photosynthesis. Vol. 48, pp. 1-44.

Llinás, M.: Metal-Polypeptide Interactions: The Conformational State of Iron Proteins. Vol. 17, pp. 135-220.

Louat, A. see Linarès, C.: Vol. 33, pp. 179-207.

Luchinat, C. see Banci, L.: Vol. 72, pp. 113-136.

Luchinat, C. see Bertini, I.: Vol. 48, pp. 45-91.

Luchinat, C. see Bertini, I.: Vol. 83, pp. 1-54.

Lucken, E. A. C.: Valence-Shell Expansion Studied by Radio-Frequency Spectroscopy. Vol. 6, pp. 1-29.

Ludi, A., Güdel, H. U.: Structural Chemistry of Polynuclear Transition Metal Cyanides. Vol. 14, pp. 1-21.

Lutz, H. D.: Bonding and Structure of Water Molecules in Solid Hydrates. Correlation of Spectroscopic

and Structural Data. Vol. 69, pp. 125.
Lutz, H. D.: Hydroxide Ions in Condensed Materials - Correlation of Spectroscopy and Structural Data. Vol. 82, pp. 85-104.

Maaskant, W. J. A.: On Helices Resulting from a Cooperative Jahn-Teller Effect in Hexagoal Perovskites. Vol. 83, pp. 55-88.
Maggiora, G. M., Ingraham, L. L.: Chlorophyll Triplet States, Vol. 2, pp. 126-159.
Magyar, B.: Salzebullioskopie III. Vol. 14, pp. 111-140.
Makovicky, E., Hyde, B. G.: Non-Commensurate (Misfit) Layer Structures. Vol. 46, pp. 101-170.
Manes, L. see Fournier, J. M.: Vol. 59/60, pp. 1-56.
Manes, L., Benedict, U.: Structural and Thermodynamic Properties of Actinide Solids and Their Relation to Bonding. Vol. 59/60, pp. 75-125.
Mann, S.: Mineralization in Biological Systems. Vol. 54, pp. 125-174.
March, N. H.: The Ground-State Energy of Atomic and Molecular Ions and Its Variation with the Number of Electrons. Vol. 80, pp. 71-86.
March, N. H.: Semiclassical Theory of Atoms and Ions in Intense External Fields. Vol. 86, pp. 63-96
Martinsen, J. see Ibers, J. A.: Vol. 50, pp. 1-55.
Mason, S. F: The Ligand Polarization Model for the Spectra of Metal Complexes: The Dynamic Coupling Transition Probabilities. Vol. 39, pp. 43-81.
Massey, V. see Hemmerich, P.: Vol. 48, pp. 93-124.
Mathey, F., Fischer, J., Nelson, J. H.: Complexing Modes of the Phosphole Moiety. Vol. 55, pp. 153-201.
Mauk, A. G.: Electron Transfer in Genetically Engineered Proteins. The Cytochrome *c* Paradigm. Vol. 75, pp. 131-158.
Mayer, U. see Gutman, V.: Vol. 10, pp. 127-151.
Mayer, U. see Gutman, V.: Vol. 15, pp. 141-166.
Mayer, H. see Gutman, V.: Vol. 31, pp. 49-66.
Mayer, U., Gutmann, V.: Phenomenological Approach to Cation-Solvent Interactions. Vol. 12, pp. 113-140.
Mazumdar, S., Mitra, S.: Biomimetic Chemistry of Hemes Inside Aqueous Micelles. Vol. 81, pp. 115-145.
McGrady, J. E. see Mingos, D. M. P.: Vol. 79, pp. 1-54.
McLendon, G.: Control of Biological Electron Transport via Molecular Recognition and Binding: The „Velcro" Model. Vol. 75, pp. 159-174.
McLeod, G. C. see Kustin, K.: Vol. 53, pp. 137-158.
Mehrotra, R. C. see Jain, V. K.: Vol. 52, pp. 147-196.
Mehrotra, R. C.: Present Status and Future Potential of the Sol-Gel Process. Vol. 77, pp. 1-36.
Meier, P. C. see Simon, W.: Vol. 16, pp. 113-160.
Melnik, M. see Valach, F.: Vol. 55, pp. 101-151.
Michel, H. see Hemmerich, P.: Vol. 48, pp. 93-124.
Mildvan, A. S., Grisham, C. M.: The Role of Divalent Cations in the Mechanism of Enzyme Catalyzed Phosphoryl and Nucleotidyl. Vol. 20, pp. 1-21.
Mingos, D. M. P., Hawes, J. C.: Complementary Spherical Electron Densitiy Model. Vol. 63, pp. 1-63.
Mingos, D. M. P., Johnston, R. L.: Theoretical Models of Cluster Bonding. Vol. 68, pp. 29-87.
Mingos, D. M. P., McGrady, J. E., Rohl, A. L.: Moments of Inertia in Cluster and Coordination Compounds. Vol. 79, pp. 1-54.
Mingos, D. M. P., Zhenyang, L.: Hybridization Schemes for Coordination and Organometallic Compounds, Vol. 72, pp. 73-112.
Mingos, D. M. P., Zhenyang, L.: Non-Bonding Orbitals in Coordination Hydrocarbon and Cluster Compounds. Vol. 71, pp. 1-56.
Mioduski, T. see Fidelis, I. K.: Vol. 47, pp. 27-51.

Mitchell, P. R. see Gillard, R. D.: Vol. 7, pp. 46-86.
Mitra, S. see Mazumdar, S.: Vol. 81, pp. 115-145.
Moody, D. C. see Ryan, R. R.: Vol. 46, pp. 47-100.
Moreau-Colin, M. L.: Electronic Spectra and Structural Properties of Complex Tetracyanides of Platinum, Palladium and Nickel. Vol. 10, pp. 167-190.
Morf, W. E. see Simon, W.: Vol. 16, pp. 113-160.
Morgan, B., Dophin, D.: Synthesis and Structure of Biometric Porphyrins. Vol. 64, pp. 115-204.
Moritz, H. G. see Keppler, B. K.: Vol. 78, pp. 97-128.
Morris, D. F. C.: An Appendix to Structure and Bonding. Vol. 4 (1968). Vol. 6, pp. 157-159.
Morris, D. F. C.: Ionic Radii and Enthalpies of Hydration of Ions. Vol. 4, pp. 63-82.
Mortensen, O. S.: A Noncommuting-Generator Approach to Molecular Symmetry. Vol. 68, pp. 1-28.
Mortier, J. W.: Electronegativity Equalization and its Applications. Vol. 66, pp. 125-143.
Mortier, W. J. see Baekelandt, B. G.: Vol. 80, pp. 187-228.
Moura, I. see Xavier, A. V.: Vol. 43, pp. 187-213.
Moura, J. J. G. see Xavier, A. V.: Vol. 43, pp. 187-213.
Mullay, J. J.: Estimation of Atomic and Group Electronegativities. Vol. 66, pp. 1-25.
Müller, A., Baran, E. J., Carter, R. O.: Vibrational Spectra of Oxo-, Thio-, and Selenometallates of Transition Elements in the Solid State. Vol. 26, pp. 81-139.
Müller, A., Diemann, E., Jørgensen, C. K.: Electronic Spectra of Tetrahedral Oxo, Thio and Seleno Complexes. Formed by Elements of the Beginning of the Transition Groups. Vol. 14, pp. 23-47.
Müller, U.: Strukturchemie der Azide. Vol. 14, pp. 141-172.
Müller, W., Spirlet, J.-C.: The Preparation of High Purity Actinide Metals and Compounds. Vol. 59/60, pp. 57-73.
Murray, J. S. see Politzer, P.: Vol. 80, pp. 101-114.
Murrell, J. M.: The Potential Energy Surfaces of Polyatomic Molecules. Vol. 32, pp. 93-146.

Naegele, J. R., Ghijsen, J.: Localization and Hybridization of 5f States in the Metallic and Ionic Bond as Investigated by Photoelectron Spectroscopy. Vol. 59/60, pp. 197-262.
Nag, K., Bose, S. N.: Chemistry of Tetra- and Pentavalent Chromium. Vol. 63, pp. 153-197.
Nalewajski, R. F.: The Hardness Based Molecular Charge Sensitivities and Their Use in the Theory of Chemical Reactivity. Vol. 80, pp. 115-186.
Natan, M. J. see Hoffman, B. M.: Vol. 75, pp. 85-108.
Neilands, J. B. see Liu, A.: Vol. 58, pp. 97-106.
Neilands, J. B. see Chimiak, A.: Vol. 58, pp. 89-96.
Neilands, J. B.: Evolution of Biological Iron Binding Centers. Vol. 11, pp. 145-170.
Neilands, J. B.: Methodology of Siderophores. Vol. 58, pp. 1-24.
Neilands, J. B.: Naturally Occurring Non-Porphyrin Iron Compounds. Vol. 1, pp. 59-108.
Neilands, J. P. see Cheh, A. M.: Vol. 29, pp. 123-169.
Nelson, J. H. see Mathey, F.: Vol. 55, pp. 153-201.
Nieboer, E.: The Lanthanide Ions as Structural Probes in Biological and Model Systems. Vol. 22, pp. 1-47.
Nocek, J. M. see Hoffman, B. M.: Vol. 75, pp. 85-108.
Nomoto, K. see Sugiura, Y.: Vol. 58, pp. 107-135.
Novack, A.: Hydrogen Bonding in Solids. Correlation of Spectroscopic and Crystallographic Data. Vol. 18, pp. 177-216.
Nultsch, W., Häder, D.-P.: Light Perception and Sensory Transduction in Photosynthetic Prokaryotes. Vol. 41, pp. 111-139.
Nyholm, R. S. see Hall, D. I.: Vol. 15, pp. 3-51.

O'Keeffe, M. see Brese, N. E.: Vol. 79, pp. 307-378.
O'Keeffe, M., Hyde, B. G.: An Alternative Approach to Non-Molecular Crystal Structures with Emphasis on the Arrangements of Cations. Vol. 61, pp. 77-144.
O'Keeffe, M.: The Prediction and Interpretation of Bond Lengths in Crystals. Vol. 71, pp. 161-190.
Odom, J. D.: Selenium Biochemistry. Chemical and Physical Studies. Vol. 54, pp. 1-26.
Oehme, I. see Wolfbeis, O. S.: Vol. 85, pp. 51-98.
Oelkrug, D.: Absorption Spectra and Ligand Field Parameters of Tetragonal *3d*-Transition Metal Fluorides. Vol. 9, pp. 1-26.
Oosterhuis, W. T.: The Electronic State of Iron in Some Natural Iron Compounds: Determination by Mössbauer and ESR Spectroscopy. Vol. 20, pp. 59-99.
Orchin, M., Bollinger, D. M.: Hydrogen-Deuterium Exchange in Aromatic Compounds. Vol. 23, pp. 167-193.
Ozaki, Y. see Kitagawa, T.: Vol. 64, pp. 71-114.

Pace, L. J. see Ibers, J. A.: Vol. 50, pp. 1-55.
Padhye, S. B. see West, D. C.: Vol. 76, pp. 1-50.
Paolesse, R. see Licoccia, S.: Vol. 84, pp. 71-134.
Parr, R. G. see Chattaraj, P. K.: Vol. 80, pp. 11-26.
Patil, S. K. see Ramakrishna, V. V.: Vol. 56, pp. 35-90.
Peacock, R. D.: The Intensities of Lanthanide $f \leftrightarrow f$ Transitions. Vol. 22, 83-122.
Pearson, R. G.: Chemical Hardness – An Historial Introduction. Vol. 80, pp. 1-10.
Pelikán, P. see Boca, R.: Vol. 71, pp. 57-97.
Penfield, K. W. see Solomon, E. I.: Vol. 53, pp. 1-56.
Penneman, R. A., Ryan, R. R., Rosenzweig, A.: Structural Systematics in Actinide Fluoride Complexes. Vol. 13, pp. 1-52.
Perlman, M. L. see Watson, R. E.: Vol. 24, pp. 83-132.
Politzer, P., Murray, J. S., Grice, M. E.: Charge Capacities and Shell Structures of Atoms. Vol. 80, pp. 101-114.
Pouchard, M. see Grenier, J. C.: Vol. 47, pp. 1-25.
Powell, R. C., Blasse, G.: Energy Transfer in Concentrated Systems. Vol. 42, pp. 43-96.

Que, Jr., L.: Non-Heme Iron Dioxygenases. Structure and Mechanism. Vol. 40, pp. 39-72.

Rakke, T. see Kjekshus, A.: Vol. 19, pp. 45-83.
Rakke, T. see Kjekshus, A.: Vol. 19, pp. 85-104.
Ramakrishna, V. V., Patil, S. K.: Synergic Extraction of Actinides. Vol. 56, pp. 35-90.
Ramakrishnan, L. see Koppikar, D. K.: Vol. 34, pp. 135-213.
Rao, V. U. S. see Wallace, W. E.: Vol. 33, pp. 1-55.
Raphael, A. L. see Therien, M. J.: Vol. 75, pp. 109-130.
Ravikanth, M., Chandrashekar, T. K.: Nonplanar Porphyrins and Their Biological Relevance: Ground and Excited State Dynamics. Vol. 82, pp. 105-188.
Rawle, S. C. see Cooper, S. R.: Vol. 72, pp. 1-72.
Raymond, K. N. see Baker, E. C.: Vol. 25, pp. 21-66.
Raymond, K. N., Smith, W. L.: Actinide-Specific Sequestering Agents and Decontamination Applications. Vol. 43, pp. 159-186.
Reedijk, J., Fichtinger-Schepman, A. M. J., Oosterom, A. T. van, Putte, P. van de: Platinum Amine Coordination Compounds as Anti-Tumour Drugs. Molecular Aspects of the Mechanism of Action. Vol. 67, pp.53-89.
Rein, M. see Schultz, H.: Vol. 74, pp. 41-146.
Reinen, D., Friebel, C.: Local and Cooperative Jahn-Teller Interactions in Model Structures.

Spectroscopic and Structural Evidence. Vol. 37, pp. 1-60.
Reinen, D.: Kationenverteilung zweiwertiger $3d_n$-Ionen in oxidischen Spinell-, Granat und anderen Strukturen. Vol. 7, pp. 114-154.
Reinen, D.: Ligand-Field Spectroscopy and Chemical Bonding in Cr_{3+}-Containing Oxidic Solids. Vol. 6, pp. 30-51.
Reisfeld, R. see Jørgensen, C. K.: Vol. 50, pp. 121-171.
Reisfeld, R., Jørgensen, C. K.: Excided States of Chromium(III) in Translucent Glass-Ceramics as Prospective Laser Materials. Vol. 69, pp. 63-96.
Reisfeld, R.: Laser Based on Sol-Gel Technology. Vol. 85, pp. 215-234.
Reisfeld, R., Jørgensen, C. K.: Luminescent Solar Concentrators for Energy Conversion. Vol. 49, pp.1-36.
Reisfeld, R.: New Materials for Nonlinear Optics. Vol. 85, pp. 99-148.
Reisfeld, R., Jørgensen, Ch. K.: Optical Properties of Colorants or Luminescent Species in Sol-Gel Glasses. Vol. 77, pp. 207-256.
Reisfeld, R.: Excited States and Energy Transfer from Donor Cations to Rare Earths in the Condensed Phase. Vol. 30, pp. 65-97.
Reisfeld, R.: Radiative and Non-Radiative Transitions of Rare Earth Ions in Glasses. Vol. 22, pp. 123-175.
Reisfeld, R.: Spectra and Energy Transfer of Rare Earths in Inorganic Glasses. Vol. 13, pp. 53-98.
Reisfeld, R. see Wolfbeis, O. S.: Vol. 85, pp. 51-98.
Reslova, S. see Thomson, A. J.: Vol. 11, pp. 1-46.
Röder, A. see Hill, H. A. O.: Vol. 8, pp. 123-151.
Rohl, A. L. see Mingos, D. M. P.: Vol. 79, pp. 1-54.
Rosenzweig, A. see Penneman, R. A.: Vol. 13, pp. 1-52.
Rüdiger, W.: Phytochrome, a Light Receptor of Plant Photomorphogenesis. Vol. 40, pp. 101-140.
Ruedenberg, K. see Hoffmann, D. K.: Vol. 33, pp. 57-96.
Runov, V. K. see Golovina, A. P.: Vol. 47, pp. 53-119.
Russo, V. E. A., Galland, P.: Sensory Physiology of *Phycomyces Blakesleeanus*. Vol. 41, pp. 71-110.
Ryan, R. R. see Penneman, R. A.: Vol. 13, pp. 1-52.
Ryan, R. R., Kubas, G. J., Moody, D. C., Eller, P. G.: Structure and Bonding of Transition Metal-Sulfur Dioxide Complexes. Vol. 46, pp. 47-100.

Sadler, P. J. see Berners-Price, S. J.: Vol. 70, pp. 27-102.
Sadler, P. J. see Dubhghaill, O. M. N: Vol. 78, pp. 129-190.
Sadler, P. J.: The Biological Chemistry of Gold: A Metallo-Drug and Heavy-Atom Label with Variable Valency. Vol. 29, pp. 171-214.
Sakka, S., Yoko, T.: Sol-Gel-Derived Coating Films and Applications. Vol. 77, pp. 89-118.
Sakka, S.: Sol-Gel Coating Films for Optical and Electronic Application. Vol. 85, pp. 1-50.
Saltman, P. see Spiro, G.: Vol. 6, pp. 116-156.
Sando, G. N. see Hogenkamp, H. P. C.: Vol. 20, pp. 23-58.
Sankar, S. G. see Wallace, W. E.: Vol. 33, pp. 1-55.
Schäffer, C. E. see Harnung, S. E.: Vol. 12, pp. 201-255.
Schäffer, C. E. see Harnung, S. E.: Vol. 12, pp. 257-295.
Schäffer, C. E.: A Perturbation Representation of Weak Covalent Bonding. Vol. 5, pp. 68-95.
Schäffer, C. E.: Two Symmetry Parameterizations of the Angular-Overlap Model of the Ligand-Field. Relation to the Crystal-Field Model. Vol. 14, pp. 69-110.
Scheidt, W. R., Lee, Y. J.: Recent Advances in the Stereochemistry of Metallotetrapyrroles. Vol. 64, pp. 1-70.
Schläpfer, C. W. see Daul, C.: Vol. 36, pp. 129-171.
Schmelcher, P. S., Cederbaum, L. S.: Two Interacting Charged Particles in Strong Static Fields: A

Variety of Two-Body Phenomena. Vol. 86, pp. 27-62

Schmid, G.: Developments in Transition Metal Cluster Chemistry. The Way to Large Clusters. Vol. 62, pp. 51-85.

Schmidt, H.: Thin Films, the Chemical Processing up to Gelation. Vol. 77, pp. 115-152.

Schmidt, P. C. see Sen, K. D.: Vol. 66, pp. 99-123.

Schmidt, P. C.: Electronic Structure of Intermetallic B 32 Type Zintl Phases. Vol. 65, pp. 91-133.

Schmidt, W.: Physiological Bluelight Reception. Vol. 41, pp. 1-44.

Schmidtke, H.-H., Degen, J.: A Dynamic Ligand Field Theory for Vibronic Structures Rationalizing Electronic Spectra of Transition Metal Complex Compounds. Vol. 71, pp. 99-124.

Schneider, W.: Kinetics and Mechanism of Metalloporphyrin Formation. Vol. 23, pp. 123-166.

Schoonheydt, R. A. see Baekelandt, B. G.: Vol. 80, pp. 187-228.

Schretzmann, P. see Bayer, E.: Vol. 2, pp. 181-250.

Schubert, K.: The Two-Correlations Model, a Valence Model for Metallic Phases. Vol. 33, pp. 139-177.

Schug, C. see Hemmerich, P.: Vol. 48, pp. 93-124.

Schultz, H., Lehmann, H., Rein, M., Hanack, M.: Phthalocyaninatometal and Related Complexes with Special Electrical and Optical Properties. Vol. 74, pp. 41-146.

Schutte, C. J. H.: The Ab-Initio Calculation of Molecular Vibrational Frequencies and Force Constants. Vol. 9, pp. 213-263.

Schweiger, A.: Electron Nuclear Double Resonance of Transition Metal Complexes with Organic Ligands. Vol. 51, pp. 1-122.

Scozzafava, A. see Bertini, I.: Vol. 48, pp. 45-91.

Sen, K. D., Böhm, M. C., Schmidt, P. C.: Electronegativity of Atoms and Molecular Fragments. Vol. 66, pp. 99-123.

Sen, K.: Isoelectronic Changes in Energy, Electronegativity, and Hardness in Atoms via the Calculations of $\langle r^{-1} \rangle$. Vol. 80, pp. 87-100.

Shamir, J.: Polyhalogen Cations. Vol. 37, pp. 141-210.

Shannon, R. D., Vincent, H.: Relationship Between Covalency, Interatomic Distances, and Magnetic Properties in Halides and Chalcogenides. Vol. 19, pp. 1-43.

Shihada, A.-F. see Dehnicke, K.: Vol. 28, pp. 51-82.

Shriver, D. F.: The Ambident Nature of Cyanide. Vol. 1, pp. 32-58.

Siegel, F. L.: Calcium-Binding Proteins. Vol. 17, pp. 221-268.

Šima, J.: Photochemistry of Tetrapyrrole Complexes. Vol. 84, pp. 135-194.

Simon, A.: Structure and Bonding with Alkali Metal Suboxides. Vol. 36, pp. 81-127.

Simon, W., Morf, W. E., Meier, P. Ch.: Specificity of Alkali and Alkaline Earth Cations of Synthetic and Natural Organic Complexing Agents in Membranes. Vol. 16, pp. 113-160.

Simonetta, M., Gavezzotti, A.: Extended Hückel Investigation of Reaction Mechanisms. Vol. 27, pp. 1-43.

Sinha, S. P.: A Systematic Correlation of the Properties of the f-Transition Metal Ions. Vol. 30, pp. 1-64.

Sinha, S. P.: Structure and Bonding in Highly Coordinated Lanthanide Complexes. Vol. 25, pp. 67-147.

Sivapullaiah, P. V. see Koppikar, D. K.: Vol. 34, pp. 135-213.

Sivý, P. see Valach, F.: Vol. 55, pp. 101-151.

Smit, H. H. A. see Thiel, R. C.: Vol. 81, pp. 1-40.

Smith, D. W., Williams, R. J. P.: The Spectra of Ferric Haems and Haemoproteins. Vol. 7, pp. 1-45.

Smith, D. W.: Applications of the Angular Overlap Model. Vol. 35, pp. 87-118.

Smith, D. W.: Ligand Field Splittings in Copper(II) Compounds. Vol. 12, pp. 49-112.

Smith, P. D. see Buchler, J. W.: Vol. 34, pp. 79-134.

Smith, T. see Livorness, J.: Vol. 48, pp. 1-44.

Smith, W. L. see Raymond, K. N.: Vol. 43, pp. 159-186.

Solomon, E. I., Penfield, K. W., Wilcox, D. E.: Active Sites in Copper Proteins. An Electric Structure Overview. Vol. 53, pp. 1-56.
Somorjai, G. A., Van Hove, M. A.: Adsorbed Monolayers on Solid Surfaces. Vol. 38, pp. 1-140.
Sonawane, P. B. see West, D. C.: Vol. 76, pp. 1-50.
Soundararajan, S. see Koppikar, D. K.: Vol. 34, pp. 135-213.
Speakman, J. C.: Acid Salts of Carboxylic Acids, Crystals with some „Very Short" Hydrogen Bonds. Vol. 12, pp. 141-199.
Spirlet, J.-C. see Müller, W.: Vol. 59/60, pp. 57-73.
Spiro, G., Saltman, P.: Polynuclear Complexes of Iron and Their Biological Implications. Vol. 6, pp. 116-156.
Steggerda, J. J. see Willemse, J.: Vol. 28, pp. 83-126.
Stewart, B. see Clarke M. J.: Vol. 36, pp. 1-80.
Strohmeier, W.: Problem und Modell der homogenen Katalyse. Vol. 5, pp. 96-117.
Sugiura, Y., Nomoto, K.: Phytosiderophores - Structures and Propterties of Mugineic Acids and Their Metal Complexes. Vol. 58, pp. 107-135.
Swann, J. C. see Bray, R. C.: Vol. 11, pp. 107-144.
Sykes, A. G.: Plastocyanin and the Blue Copper Proteins. Vol. 75, pp. 175-224.

Takita, T. see Umezawa, H.: Vol. 40, pp. 73-99.
Tam, S.-C., Williams, R. J. P.: Electrostatics and Biological Systems. Vol. 63, pp. 103-151.
Teller, R., Bau, R. G.: Crystallographic Studies of Transition Metal Hydride Complexes. Vol. 44, pp. 1-82.
Therien, M. J., Chang, J., Raphael, A. L., Bowler, B. E., Gray, H. B.: Long-Range Electron Transfer in Metalloproteins. Vol. 75, pp. 109-130.
Thiel, R. C., Benfield, R. E., Zanoni, R., Smit, H. H. A., Dirken, M. W.: The Physical Properties of the Metal Cluster Compound $Au_{55}(PPh_3)_{12}Cl_6$. Vol. 81, pp. 1-40.
Thomas, K. R. see Chandrasekhar, V.: Vol. 81, pp. 41-114.
Thompson, D. W.: Structure and Bonding in Inorganic Derivatives of ß-Diketones. Vol. 9, pp. 27-47.
Thomson, A. J., Williams, R. J. P., Reslova, S.: The Chemistry of Complexes Related to *cis*-$Pt(NH_3)_2Cl_2$. An Anti-Tumor Drug. Vol. 11, pp. 1-46.
Tofield, B. C.: The Study of Covalency by Magnetic Neutron Scattering. Vol. 21, pp. 1-87.
Trautwein, A. X., Bill, E., Bominaar, E. L., Winkler, H.: Iron-Containing Proteins and Related Analogs-Complementary Mössbauer, EPR and Magnetic Susceptibility Studies. Vol. 78, pp. 1-96.
Trautwein, A.: Mössbauer-Spectroscopy on Heme Proteins. Vol. 20, pp. 101-167.
Tressaud, A., Dance, J.-M.: Relationships Between Structure and Low-Dimensional Magnetism in Fluorides. Vol. 52, pp. 87-146.
Tributsch, H.: Photoelectrochemical Energy Conversion Involving Transition Metal d-States and Intercalation of Layer Compounds. Vol. 49, pp. 127-175.
Truter, M. R.: Structures of Organic Complexes with Alkali Metal Ions. Vol. 16, pp. 71-111.

Umezawa, H., Takita, T.: The Bleomycins: Antitumor Copper-Binding Antibiotics. Vol. 40, pp.73-99.

Vahrenkamp, H.: Recent Results in the Chemistry of Transition Metal Clusters with Organic Ligands. Vol. 32, pp. 1-56.
Valach, F., Kóren, B., Sivý, P., Melník, M.: Crystal Structure Non-Rigidity of Central Atoms for Mn(II), Fe(II), Fe(III), Co(II), Co(III), Ni(II), Cu(II) and Zn(II) Complexes. Vol. 55, pp. 101-151.

Valdemoro, C. see Fraga, S.: Vol. 4, pp. 1-62.
van Bronswyk, W.: The Application of Nuclear Quadrupole Resonance Spectroscopy to the Study of Transition Metal Compounds. Vol. 7, pp. 87-113.
van de Putte, P. see Reedijk, J.: Vol. 67, pp. 53-89.
Van Hove, M. A. see Somorjai, G. A.: Vol. 38, pp. 1-140.
van Oosterom, A. T. see Reedijk, J.: Vol. 67, pp. 53-89.
Vanquickenborne, L. G. see Ceulemans, A.: Vol. 71, pp. 125-159.
Vela, A. see Gázquez, J. L.: Vol. 66, pp. 79-98.
Verkade, J. G. see Hoffmann, D. K.: Vol. 33, pp. 57-96.
Vincent, H. see Shannon, R. D.: Vol. 19, pp. 1-43.
Vogel, E. see Keppler, B. K.: Vol. 78, pp. 97-128.
von Herigonte, P.: Electron Correlation in the Seventies. Vol. 12, pp.1-47.
von Zelewsky, A. see Daul, C.: Vol. 36, pp. 129-171.
Vongerichten, H. see Keppler, B. K.: Vol. 78, pp. 97-128.

Wallace, W. E. Sankar, S. G., Rao, V. U. S.: Field Effects in Rare-Earth Intermetallic Compounds. Vol. 33, pp. 1-55.
Wallin, S. A. see Hoffman, B. M.: Vol. 75, pp. 85-108.
Walton, R. A. see Cotton, F. A.: Vol. 62, pp. 1-49.
Warren, K. D. see Allen, G. C.: Vol. 19, pp. 105-165.
Warren, K. D. see Allen, G. C.: Vol. 9, pp. 49-138.
Warren, K. D. see Clack, D. W.: Vol. 39, pp. 1-141.
Warren, K. D.: Calculations of the Jahn-Teller Coupling Constants for d_x Systems in Octahedral Symmetry via the Angular Overlap Model. Vol. 57, pp. 119-145.
Warren, K. D.: Ligand Field Theory of f-Orbital Sandwich Complexes. Vol. 33, pp. 97-137.
Warren, K. D.: Ligand Field Theory of Metal Sandwich Complexes. Vol. 27, pp. 45-159.
Watson, R. E., Perlman, M. L.: X-Ray Photoelectron Spectroscopy. Application to Metals and Alloys. Vol. 24, pp. 83-132.
Weakley, T. J. R.: Some Aspects of the Heteropolymolybdates and Heteropolytungstates. Vol. 18, pp. 131-176.
Weissbluth, M.: The Physics of Hemoglobin. Vol. 2, pp. 1-125.
Wendin, G.: Breakdown of the One-Electron Pictures in Photoelectron Spectra. Vol. 45, pp. 1-130.
Wertheim, G. K. see Campagna, M.: Vol. 30, pp. 99-140.
Weser, U.: Chemistry and Structure of some Borate Polyol Compounds. Vol. 2, pp. 160-180.
Weser, U.: Reaction of some Transition Metals with Nucleic Acids and Their Constituents. Vol. 5, pp. 41-67.
Weser, U.: Redox Reactions of Sulphur-Containing Amino-Acid Residues in Proteins and Metalloproteins, an XPS Study. Vol. 61, pp. 145-160.
Weser, U.: Structural Aspects and Biochemical Function of Erythrocuprein. Vol. 17, pp. 1-65.
West, D. C., Padhye, S. B., Sonawane, P. B.: Structural and Physical Correlations in the Biological Properties of Transitions Metal Heterocyclic Thiosemicarbazone and S-alkyldithiocarbazate Complexes. Vol. 76, pp, 1-50.
Wetterhahn, K. E. see Connett, P. H.: Vol. 54, pp. 93-124.
Wilcox, D. E. see Solomon, E. I.: Vol. 53, pp. 1-56.
Willemse, J., Cras, J. A., Steggerda, J. J., Keijzers, C. P.: Dithiocarbamates of Transition Group Elements in „Unusual“ Oxidation State. Vol. 28, pp. 83-126.
Williams, A. F. see Gubelmann, M. H.: Vol. 55, pp. 1-65.
Williams, R. J. P. see Fraústo da Silva, J. J. R.: Vol. 29, pp. 67-121.
Williams, R. J. P. see Hill, H. A. O.: Vol. 8, pp. 123-151.
Williams, R. J. P. see Smith, D. W.: Vol. 7, pp. 1-45.
Williams, R. J. P. see Tam, S.-C.: Vol. 63, pp. 103-151.

Williams, R. J. P. see Thomson, A. J.: Vol. 11, pp. 1-46.
Williams, R. J. P., Hale, J. D.: Professor Sir Ronald Nyholm. Vol. 15, pp. 1 and 2.
Williams, R. J. P., Hale, J. D.: The Classification of Acceptors and Donors in Inorganic Reactions. Vol. 1, pp. 249-281.
Williams, R. J. P.: The Chemistry of Lanthanide Ions in Solution and in Biological Systems. Vol. 50, pp. 79-119.
Wilson, J. A.: A Generalized Configuration-Dependent Band Model for Lanthanide Compounds and Conditions for Interconfiguration Fluctuations. Vol. 32, pp. 57-91.
Winkler, H. see Trautwein, A. X.: Vol. 78, pp. 1-96.
Winkler, R.: Kinetics and Mechanism of Alkali Ion Complex Formation in Solution. Vol. 10, pp. 1-24.
Wolfbeis, O. S., Reisfeld, R., Oehme, I.: Sol-Gels and Chemical Sensors. Vol. 85, pp. 51-98
Wood, J. M., Brown, D. G.: The Chemistry of Vitamin B12-Enzymes. Vol. 11, pp. 47-105.
Woolley, R. G. see Gerloch, M.: Vol. 46, pp. 1-46.
Woolley, R. G.: Natural Optical Acitivity and the Molecular Hypothesis. Vol. 52, pp. 1-35.
Wüthrich, K.: Structural Studies of Hemes and Hemoproteins by Nuclear Magnetic Resonance Spectroscopy. Vol. 8, pp. 53-121.

Xavier, A. V., Moura, J. J. G., Moura, I.: Novel Structures in Iron-Sulfur Proteins. Vol. 43, pp. 187-213.

Yersin, H. see Gliemann, G.: Vol. 62, pp. 87-153.
Yoko, T. see Sakka, S.: Vol. 77, pp. 89-118.

Zanchini, C. see Banci L.: Vol. 52, pp. 37-86.
Zanello, P.: Stereochemical Aspects Associated with the Redox Behaviour of Heterometal Carbonyl Clusters. Vol. 79, pp. 101-214.
Zanoni, R. see Thiel, R. C.: Vol. 81, pp. 1-40.
Zhenyang, L. see Mingos, D. M. P.: Vol. 71, pp. 1-56.
Zhenyang, L. see Mingos, D. M. P.: Vol. 72, pp. 73-112.
Zorov, N. B. see Golovina, A. P.: Vol. 47, pp. 53-119.
Zumft, W. G.: The Molecular Basis of Biological Dinitrogen Fixation. Vol. 29, pp. 1-65.

Subject Index

above threshold ionization 2
action integral 104
adiabatic connection 133
angular frequency 87
angular momentum states 73, 115
anharmonicity 40, 168
anions 88
argon 24
argon-ion laser 3
astrophysical magnetic fields 56
atomic ions 74, 79, 87
atomic needles 89
atomic units 151
axial symmetry 81
azimuthal quantum number 71, 129

barrier penetration 73
Bessel function 139
bifurcation 122
binding energy 44
Bloch equation 66, 80, 81
bond hardening 3, 15, 25
bond softening 3, 15, 19
bosonized 76

canonical momentum 30, 43, 50
canonical density matrix 65, 66, 82, 86, 89, 91
celestial mechanics 62
center of mass 28, 42, 60
chaos 98
chaotic scattering 99, 100, 103, 114, 120, 122
charge density 95
chemical potential 79
chemical reactions 3
CH^+ 6
Circular polarization 20, 99, 112, 129
classical chaos 98
classical trajectory 36
CO 2, 11
coherent atomic excitation 172
collective motion 52
completeness condition 66
confined atoms 79
correlation coefficient 10
Coulomb explosion 2, 15
Coulomb potential 30, 37, 49, 100, 104, 108
Coulomb singularity 37, 82
covariance mapping 9
crossed fields 30, 37, 60
current density 65, 88, 91, 92
cyclotron motion 49

degrees of freedom 2, 29
delta function 66
density amplitude 75, 88
density functional theory 95
density matrix 65, 66, 76
diamagnetic term 32, 54
diatomic molecules 13
differential equation 90
diffusion 42, 44
diffusive growth 98
dipole matrix elements 53, 126
dipole moment 61, 126
dimensional perturbation theory 70
dimensional scaling 95
Dirac density matrix 65
dissociation energy 89
drift velocity 30

effective potential 90, 156, 157
eigenfunctions 34, 66
eigenvalues 34, 102
electric field propagator 87
electrodynamics 62
electronic excitation 55
elevated temperature 36
energy gap 39
energy transfer 29, 52, 62
equation of motion 98
exchange-correlation potential 75

excitation energy 98
excited state 98
excitons 45
expansion parameter 71

Fabry-Perot cavity 129
femtosecond laser 4
Fermi-Dirac function 66
Fermi-Dirac statistics 84, 86
Feynman propagator 65, 89
fine structure constant 80
fixed nuclei approximation 28
Floquet approach 127, 137, 140
Fourier transform 35, 102
fractal dimension 102, 110
free electrons 66, 151
free particles 82, 120
free trajectories 100

galactic magnetic fields 65
gauge 91
gauge dependence 31
gauge invariance 31
gaussian fluctuations 72
generic model 99
gradient expansion 94
Green function 86, 90
group velocity 5

Hamiltonian 29, 30, 33, 42, 50, 52, 60, 65, 109, 151
harmonic confinement 83
harmonic force 81
harmonic generation 2, 168
harmonic oscillator 81
harmonic potential 34, 80
H atom 28, 37, 39, 44, 48, 60, 72, 89, 98, 99, 104, 111, 126, 130
He^+ ion 54, 57, 59
Hermite polynomials 36
High harmonic generation 2, 3, 56, 170
hydrocarbons 6
hydrogen molecule ion 17, 22, 87

$\mathbf{I}_2$ 13, 21
inertial confinement 25
inhomogeneous electron gas 95
integrability 98
internal motion 29
internuclear separation 88
ionization threshold 28

K 139
Kepler ellipses 57
kinetic energy 11, 21, 25, 31, 51, 58
kinetic momentum 43
Kolmogorov-Arnold-Moser tori 99, 103, 105, 112

laboratory coordinates 31
Landau energy 52, 53
Landau limit 71
Landau-Zener transition 132, 133, 135, 137, 144
Langevin equation 44
Larmor frequency 57
Larmor radius 57
laser field 3, 98
laser flux 84, 85
laser-induced stabilization 25
Levi-Civita tensor 50
Liapunov exponents 46, 103, 119, 120
linear polarization 17, 99
linear response 90
LS-coupling 158

magnetic levitation 6
Maxwell-Boltzmann statistics 84
methane 24
microwave field 98, 126
molecular ions 25, 65, 87
momentum operator 34, 65
motional Stark term 32
multielectron dissociative ionization 3
multiphoton ionization 80
multiphoton processes 26, 150

Na 128, 129
Nd 23
Neon 150
neutron stars 28
Newton equations 51, 56
N_2 3, 6, 9, 15, 24
non-degenerate plasma 80, 85
nonlinear dynamics 99, 120
nonlinear equation 93
nonlinear phenomena 2

nonlocal relation 92
non-relativistic theory 76, 104, 151
nuclei 28, 82

orthonormality 66
oscillator frequency 80

Pauli potential 75
phase factor 88
phase function 92
phase space 43, 49, 57, 100, 107, 112
photoelectron spectrum 2, 3
plasma density 80
plasmas 65
polarizability 24
positronium 42
principal quantum number 54, 104
pseudomomentum 30, 31, 37, 44

quantum chaos 122
quantum defect 129
quantum fluctuations 73
quantum Hall effect 28
quantum oscillator 33

Rabi frequency 137, 139, 142
relativistic kinetic energy 79
Rydberg states 28, 37, 45, 62, 70, 126, 127, 139

Schrödinger equation 34, 142, 158, 170
self-consistent field 74
self ionization 29, 60, 62
self-similar 102
self-stabilization 29, 57
semiclassical theory 65, 69, 102
signal-to-noise 14
Slater determinant 154
Slater-Kohn-Sham potential 75
Slater sum 70, 74, 89, 92
space charge 5
spherical harmonics 152
spin-orbit coupling 157
Stark term 31, 130
stroboscopic map 105
survival probability 102, 103, 109, 121
symmetric gauge 33, 91

terrestrial chemistry 76, 88
Thomas-Fermi equation 65
Thomas-Fermi theory 65, 83, 86
threshold field strength 98
Ti 5, 150
time-dependent theory 58, 89, 151, 159, 170
trajectory 25, 36, 43, 47, 100, 110, 118
translation group 30
tunneling 22, 24, 83
turbomolecular pumps 5
two-electron atoms 96, 114, 147

ultrafast phenomena 25

vacuum techniques 5
vector potential 30, 65
virial 77
von Weizsäcker equation 75

wave functions 40, 41 153, 159, 170
white dwarfs 28

Xe 5, 20, 24, 156, 158, 161, 164, 169
X radiation 3

Young's two slit experiment 135

Zeeman energy 54

Printing: Saladruck, Berlin
Binding: Buchbinderei Lüderitz & Bauer, Berlin